HIGH-TECH LEGO® PROJECTS

HIGH-TECH LEGO® PROJECTS

16 rule-breaking inventions

grady koch

Printed in USA

23 22 21 20   1 2 3 4 5 6 7 8 9

ISBN-13: 978-1-7185-0025-9 (print)
ISBN-13: 978-1-7185-0026-6 (ebook)

Publisher: William Pollock
Executive Editor: Barbara Yien
Production Manager: Laurel Chun
Production Editors: Kassie Andreadis and Meg Sneeringer
Interior and Cover Design: Octopod Studios
Developmental Editor: Frances Saux
Technical Reviewer: Xander Soldaat
Copyeditor: Sharon Wilkey
Compositor: Kim Scott, Bumpy Design
Proofreader: Paula L. Fleming
Indexer: JoAnne Burek

For information on distribution, translations, or bulk sales, please contact No Starch Press, Inc. directly:
No Starch Press, Inc.
245 8th Street, San Francisco, CA 94103
phone: 1.415.863.9900; info@nostarch.com; www.nostarch.com

The Library of Congress has cataloged the first edition as follows:
Names: Koch, Grady, author.
Title: High-tech LEGO projects: 16 rule-breaking inventions / Grady Koch.
Description: San Francisco : No Starch Press, [2021] | Includes index.
Identifiers: LCCN 2020006808 (print) | LCCN 2020006809 (ebook) | ISBN
    9781718500259 (paperback) | ISBN 9781718500266 (ebook)
Subjects: LCSH: LEGO Mindstorms toys. | Mechatronics--Design and
    construction.
Classification: LCC TJ163.12 .K63 2020 (print) | LCC TJ163.12 (ebook) |
    DDC 621--dc23
LC record available at https://lccn.loc.gov/2020006808
LC ebook record available at https://lccn.loc.gov/2020006809

To Melissa, with thanks for tolerating all the
LEGO pieces strewn throughout our home.

about the author

Grady Koch is the creator of hightechlego.com. He has a Ph.D. in electrical engineering, 30 years of experience as a research engineer at NASA Langley Research Center, and three patents in the field of LiDAR for studying the atmosphere. He has written or contributed to over 175 conference and seminar papers, including 13 peer-reviewed journal papers as first author. Throughout his career, Koch has worked with student interns and learned how to bridge the gap between a student's academic instruction and laboratory practice.

about the tech reviewer

Xander Soldaat is a former Mindstorms Community Partner for LEGO MINDSTORMS. He was an IT infrastructure architect and engineer for 18 years before becoming a full-time software developer, first for Robomatter and VEX Robotics, and now as an R&D Engineer for an embedded Wi-Fi solutions provider. In his spare time he likes to tinker with robots, 3D printing, and home-built retro-computers.

acknowledgments

My appreciation is given to The LEGO Group, of course. I find building with LEGO to be gratifying from both an artistic and engineering perspective. I've been playing with LEGO bricks for decades, and I'm still intrigued by the way they stick together with just the right amount of friction.

I've also enjoyed the work of several other companies in my LEGO-building pursuits, including Mindsensors (*mindsensors.com*) and Dexter Industries (*dexterindustries .com*) with their specialized LEGO-compatible sensors. The Grove product line made by Seeed (*seeedstudio.com*) has also been a source of inspiration.

I made the assembly instructions throughout this book with LDraw tools (*ldraw.org*). LDraw, originally built by James Jessiman, is a powerful, open source computer-aided design (CAD) for LEGO creations that has been expanded on by many dedicated people. There are several interfaces and implementations of the LDraw foundation, of which I mostly used Mike's LEGO CAD (*mlcad.lm-software.com*) developed by Michael Lachmann. My appreciation is also given to the LSynth utility (*lsynth.sourceforge.net*), developed by Don Heyse and Kevin Clague, that allows for drawing cables and bendable parts. To create building instructions from CAD models, I used the LIC utility (*bugeyedmonkeys.com*), developed by Remi Gagne.

Several experts provided advice in the development of this book, including Marek Smith who identified the birds photographed in Chapter 1. I frequently visited Holly Wood's series of tutorials on learning LDraw at *holly-wood.it*. Xander Soldaat served as an excellent technical reviewer, offering many suggestions and insights for making this book better.

For years I've enjoyed the books published by No Starch Press due to their attention to detail, artistic eye, and enthusiasm for delving into diverse geeky subjects. So I have been delighted to work with No Starch Press, starting with Bill Pollock's insight that a book on highly technical LEGO applications might be a good idea. Annie Choi helped me greatly with the early formulation of the book. Frances Saux has been a wonderful editor, guiding me on a path to balancing technical content with reading enjoyment. Producing and laying out a book is an art form, as admirably demonstrated by Kassie Andreadis, Meg Sneeringer, and the No Starch Press team.

Finally, let me thank my kids, Kirsten and Elias, for providing me the excuse to buy a lot of LEGO and for the many happy hours of shared building.

brief contents

contents in detail

introduction

In this book, you'll explore LEGO MINDSTORMS and Technic with projects that teach you how to monitor for intruders, automatically photograph wildlife, experiment with lasers and light-emitting diodes, send Morse code, set up covert infrared communications, capture nocturnal insects, build your own sensors, and more.

You'll use MINDSTORMS EV3 sensors and motors and take advantage of these devices' more advanced capabilities. You'll also learn to include non-LEGO sensors and electronic components for added capability. Treat the projects in this book as springboards for new ideas and LEGO applications. One of the joys of LEGO is that building new inventions is quick, easy, and fun—and, now, practical. With MINDSTORMS EV3's computer control, software, sensors, and motors, there's no need to keep your LEGO collection confined to models or tabletop robots. Build something useful: perhaps a scientific instrument or something you can take outdoors to explore the world.

what is MINDSTORMS EV3?

MINDSTORMS is a LEGO product line that allows you to add computer control, motors, and sensors to LEGO inventions. MINDSTORMS has undergone several upgrades since its introduction in 1998, and the latest version is known as MINDSTORMS EV3. Though developed for building robots, MINDSTORMS EV3 serves as an excellent toolkit for inventing all sorts of new devices. You can use LEGO's structural building capability, combined with sensors and motors controlled through computers, to perform experiments involving astronomy, biology, communications, security, and surveillance, as you will in this book.

what is Technic?

Technic is a LEGO product line for building complex structures and machines. You can attach LEGO elements to the sides of Technic bricks, not just to the top or bottom, allowing for rugged designs. The Technic sets include mechanical elements such as gears, liftarms, pins, and axles for building machines with moving parts. These aspects of Technic are a useful complement to MINDSTORMS EV3 because they provide a way to hold and orient sensors and motors, as you'll see throughout this book. Technic inventions that don't rely on MINDSTORMS EV3 also offer interesting applications. We'll explore these in four chapters of this book.

what you'll need to build the projects in this book

Each chapter identifies all the parts you'll need to build the projects, with diagrams that include the part's picture, quantity, name, and part number. This information should help you find the parts you already have in your collection and identify any additional ones you may need to purchase.

Sometimes, combining bricks from various LEGO product lines lets you build the best projects. Most of the parts needed for this book are from the MINDSTORMS EV3 #31313 set, but some parts (identified by asterisks in the part list diagrams in each chapter) are from the Technic or LEGO System product lines. In some cases, though, I suggest an alternate design using other parts from the MINDSTORMS EV3 #31313 set.

You can easily find any LEGO part you're missing at online sites such as BrickLink (*https://www.bricklink.com/*). Simply enter the part number into the search bar, as shown in Figure 1 for a 2 × 4 brick (part number 3001).

Each chapter provides the part number to use for a search. Each part is shown as a picture and listed in the following format: 3001 2 × 4 Brick. The first number is the part number (3001), and the second number is the name of the part (a 2 × 4 brick).

Some projects include aftermarket LEGO-compatible sensors, electronic components, cameras, or tools. The text explains the requirements for using each of these non-LEGO devices, along with information about where to buy them.

Finally, in several chapters, you'll need the following:

* A soldering iron and solder

* Diagonal cutters for trimming wires

* Wire strippers

* Cyanoacrylate glue (such as Super Glue, Krazy Glue, or Gorilla Glue)

* Jumper wires

Some of the projects in this book use the LEGO MINDSTORMS EV3 Home Edition software. To be able to use this software, you will need macOS 10.14 or earlier, or Windows Vista or higher.

who this book is for

This book is best suited for someone who already has some experience with LEGO and MINDSTORMS EV3—for example, someone who has built one or two of the robots in the MINDSTORMS EV3 #31313 set. That's because these projects use the same sort of terminology, instructions, and programming environment used in those MINDSTORMS projects.

That said, each chapter offers step-by-step building instructions, along with ideas for altering the projects to make something a little different. Each chapter also walks through writing the software in the MINDSTORMS EV3 programming environment's block-based language, block by block, so you can write the code even if you're not already familiar with the environment (or programming in general). Some of the projects use algebra that is generally taught in high school.

who i am

I've always been interested in LEGO, and I love tinkering with new devices. This curiosity put me on a path to studying engineering. I eventually graduated with a PhD in electrical engineering and have a career in high-tech research. Through this journey, I've learned a lot of science and math, which I enjoy implementing in LEGO-based inventions as a

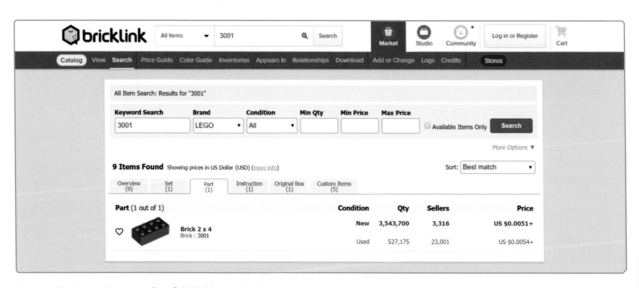

FIGURE 1: You can purchase parts from BrickLink by entering the part number in the search bar at the top of the screen.

hobby. As LEGO grew more sophisticated with the advent of MINDSTORMS, I found that building complex devices became quick, easy, and fun.

I noticed that my inventions often caught kids' attention because they were built from LEGO pieces, serving as a familiar and playful way for them to learn new science, technology, and math concepts. I wrote this book to share this enjoyment and help other people incorporate science and engineering concepts in their own LEGO pursuits with practical, step-by-step instructions.

the projects

This book features 16 advanced LEGO technology projects across 12 chapters:

Chapter 1: The Motion-Activated Critter Cam Shows you how to control a digital camera to automatically photograph animals outdoors. Take high-quality photographs of wildlife that visits your area.

Chapter 2: The LEGO-Compatible Laser Teaches you how to construct a laser that is compatible with LEGO mounting. The laser not only looks cool but also is useful for installing an alarm system, putting on a laser light show, marking a target, or making holograms.

Chapter 3: The Laser Security Fence Uses the LEGO-Compatible Laser built in Chapter 2 to build a laser-beam alarm system. Monitor your home for intruders or trespassers on your property.

Chapter 4: Morse Code Transmitters Helps you build two MINDSTORMS EV3 Morse code devices to transmit greetings or get help in emergencies. One project is a practice Morse code transmitter to help you memorize Morse code characters, and the second project is a device that lets you record and then flash a continuous, automated message.

Chapter 5: Infrared Message Transmitters Introduces you to infrared light, which is invisible to human eyes, and uses it to send hidden messages. You'll build two projects in this chapter: an infrared message transmitter and an infrared message receiver. These projects will also teach you how to control non-LEGO electronics by using MINDSTORMS EV3.

Chapter 6: The Automatic Insect Trapper Lets you easily capture outdoor nocturnal insects that are otherwise difficult to collect for study. You'll use ultraviolet light to attract insects into a collection box and then automatically close the trap to secure the insects.

Chapter 7: The Motion-Sensing Radar Connects a Doppler radar to MINDSTORMS EV3 to detect motion, even through solid objects. This project shows you how to add your own sensors to MINDSTORMS.

Chapter 8: The Tower of Eratosthenes Allows you to figure out your geographic latitude and calculate the tilt of the earth on its axis. You'll measure these parameters from the shadow of the sun cast by a custom-designed tower.

Chapter 9: Hacking LEGO Light Bricks Shows you how to create any light brick color, beyond the factory-installed options of red or yellow. Then you'll add a lens to a light brick to build a simulated laser.

Chapter 10: The Flickering Fireplace Introduces you to a programmable light-emitting diode whose color and brightness you can dynamically change. Add this lighting effect to your LEGO models, such as the fireplace you'll build in this chapter.

Chapter 11: The Laser Light Show Combines spinning mirrors with the LEGO-Compatible Laser built in Chapter 2 to project geometric patterns. You'll learn how to use the optical properties of a mirror.

Chapter 12: The Infrared Thermometer and Cannon Work with far-infrared wavelengths to measure the temperature of distant objects. You'll build a non-contact thermometer and a heat-seeking cannon that shoots LEGO studs.

online bonus projects

In addition to the projects printed in the book, you can find two more projects at *https://nostarch.com/high-tech-lego/*:

Bonus Project 1: The LEGO Car Tracker Adds a GPS sensor to MINDSTORMS. You can track the location of a car to record your own travels or those of someone you want to monitor.

Bonus Project 2: The LEGO Satellite Spotter Helps you locate satellites passing overhead with a pointer that shows you exactly where to look.

1

the motion-activated critter cam

In this chapter, you'll build a Motion-Activated Critter Cam that uses the MINDSTORMS EV3 Infrared Sensor as a proximity detector to monitor outdoor areas. You can use this surveillance system to observe animals without scaring them away. For example, say you've found bite and scratch marks all over a play fort in the backyard, as shown in Figure 1-1.

FIGURE 1-1: *An unidentified animal has been visiting a play fort at night to gnaw and scratch the wood.*

To find out what animal is gnawing or scratching the wood, you could attach a *trail camera* with built-in motion detection to a tree and leave it unattended for weeks. Hunters use trail cameras to figure out the travel and feeding patterns of large animals such as deer.

But trail cameras aren't always the best choice when you want to do the following:

Photograph smaller animals You might want to watch birds, squirrels, or raccoons, for example. Only large animals, such as deer or humans, can trigger trail cameras.

Monitor a small area Trail cameras capture a large area, such as a section of a field, rather than a precise space, such as a bird feeder or the inside of a play fort.

Use a single-lens reflex (SLR) digital camera Trail cameras have limited photograph resolution, no optical zoom, no way to change lenses, no method to set focus, and no controllable flash. SLR cameras provide better-quality photos and control.

Capture fast-moving objects Trail cameras of reasonable cost have a delay between the triggering event and the photograph.

With the surveillance system you'll build in this chapter, you'll be able to take better-quality pictures and focus on more specific areas of your surroundings. The Motion-Activated Critter Cam uses a sensor that can probe small, precise areas, and the speed of the controlling MINDSTORMS EV3 software allows you to capture animal motion more quickly than with a trail camera.

building the motion-activated critter cam

The design for the Motion-Activated Critter Cam, shown in Figure 1-2, uses the EV3 Infrared Sensor as a proximity detector.

The system consists of an infrared sensor head, an EV3 Intelligent Brick, a button pusher, and an infrared remote control for your camera.

When something triggers the proximity detector, the EV3 Intelligent Brick turns an EV3 Medium Motor, which pushes a button on a remote control for an SLR camera. This remote control sends commands to the camera to snap a picture.

what you'll need

Figures 1-3 and 1-4 show the LEGO parts you need to build the Motion-Activated Critter Cam. You can find most of these bricks in the MINDSTORMS EV3 #31313 set. The bricks marked with an asterisk in Figure 1-4 are not included in the set. The part number shown next to each part can help you locate it at sites such as BrickLink (*https://www.bricklink.com/v2/main.page*) in case you need to buy parts.

The projects in this book combine the different styles of Technic (both studded and studless), as well as LEGO System elements, to achieve the best performance with the fewest bricks. But if you prefer all-studless Technic designs, you can modify the designs in this chapter and throughout this book to your liking. For example, the locking turntable used in this project could be replaced with a Technic pin inserted into a liftarm.

You'll also need an SLR camera with a remote control. Almost all SLR cameras have a remote control available as an accessory. Over the years, Canon has made many versions of the entry-level, low-cost Rebel model, any of which will work well for this project. You might need a tripod for the camera and a platform, such as a ladder, for the Motion-Activated Critter Cam to sit on.

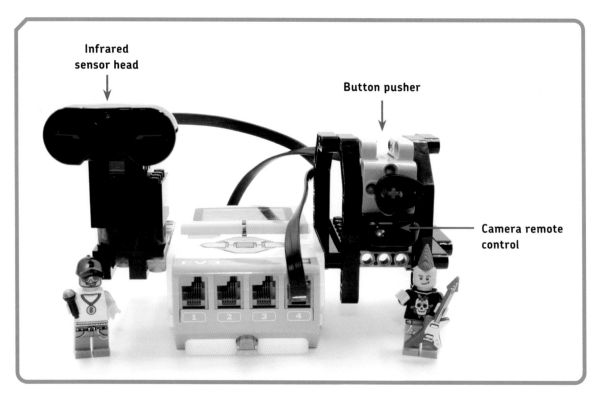

FIGURE 1-2: *The Motion-Activated Critter Cam*

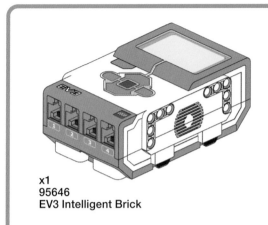

x1
95654
EV3 Infrared Sensor

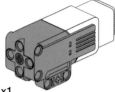

x1
99455
EV3 Medium Motor

x1
95646
EV3 Intelligent Brick

x2
55805
Connector Cable

FIGURE 1-3: EV3 parts used in the Motion-Activated Critter Cam

x1
6575
Technic Cam

x1
3032
4 x 6 Plate*

x1
64179
5 x 7 Open Center Frame Liftarm

x2
3701
1 x 4 Technic Brick with Holes*

x2
32524
1 x 7 Technic Liftarm

x1
4265c
Technic Bush 1/2 Smooth

x11
2780
Technic Pin with Friction Ridges

x1
32316
1 x 5 Technic Liftarm

x4
43093
Technic Axle Pin

x4
32009
Double Bent Liftarm

x4
3001
2 x 4 Brick*

x3
6558
Long Technic Pin with Friction Ridges

x2
2730
1 x 10 Technic Brick with Holes*

x1
87081
Locking Turntable*

x4
48989
Technic 4-Pin Connector

x1
32062
Technic Axle 2 Notched

FIGURE 1-4: Other LEGO parts used in the Motion-Activated Critter Cam

building the infrared sensor head

The infrared sensor head, shown in Figure 1-5, will detect the presence of an animal.

You'll use a MINDSTORMS EV3 sensor, typically used to help robots detect obstacles, in proximity mode. You can precisely aim the EV3 Infrared Sensor at a target location by mounting the sensor on a locking turntable. Follow the steps shown here to build it.

FIGURE 1-5: *Rotate and aim the EV3 Infrared Sensor by mounting it to a locking turntable.*

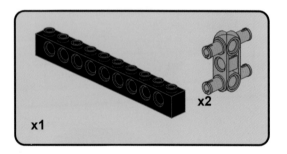

x1 x2

1

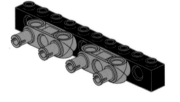

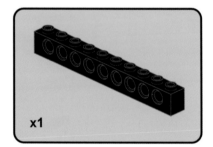

x1

2

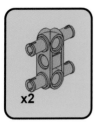

x2

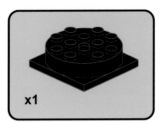

x1

3

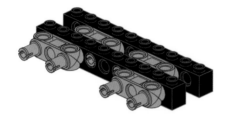

4

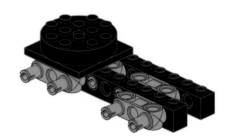

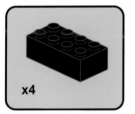

x4

5

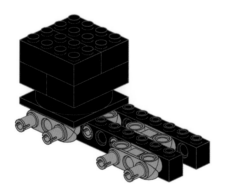

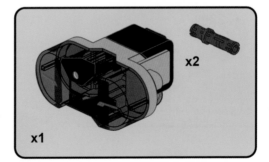

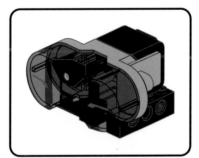

x2

x1

6

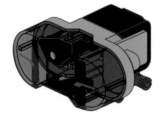

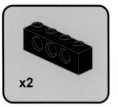

x2

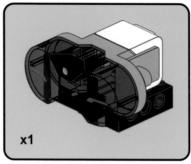

x1

7

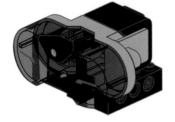

8

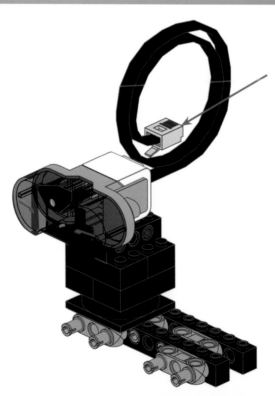

This stack of 2 × 4 bricks sets the height of the EV3 Infrared Sensor. You may want to add more to the stack when you set up the Motion-Activated Critter Cam.

x1

Connect the end of the cable to port 4 of the EV3 Intelligent Brick.

9

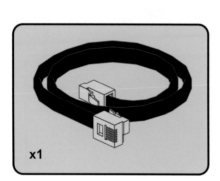

building the button pusher

The button pusher will activate the button on the camera's remote control whenever the EV3 Infrared Sensor detects motion. You'll use an EV3 Medium Motor to turn a cam, as shown in Figure 1-6.

A *cam* is an egg-shaped attachment that you can add to a rotating machine to create a small lifting or pushing action. As the motor turns, the pointy end of the cam nudges against the button to activate the remote control. Follow the instructions shown here to build it.

Cam

FIGURE 1-6: The button pusher uses a cam to press against a remote control.

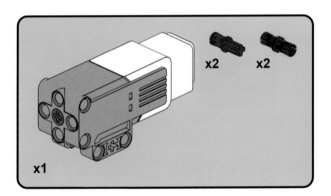

x2   x2

x1

1

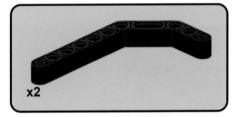

x2

2

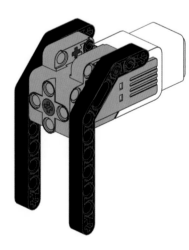

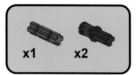

3

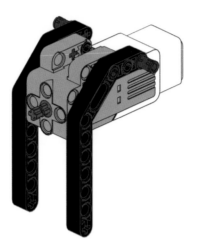

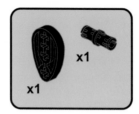

4

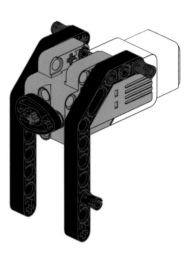

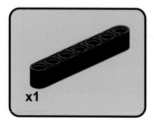

5

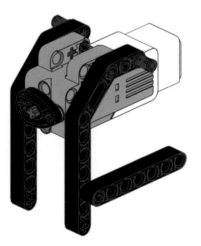

6

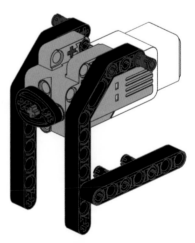

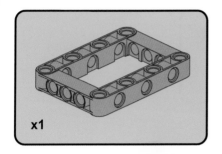

x1

7

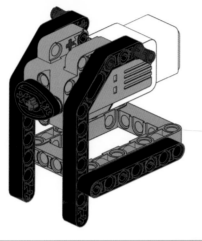

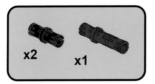

x2 **x1**

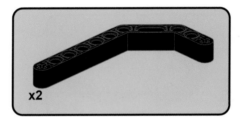

8

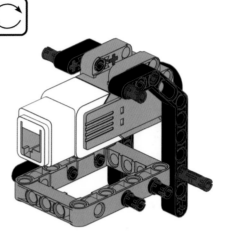

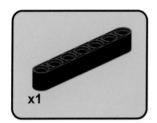

x1

9

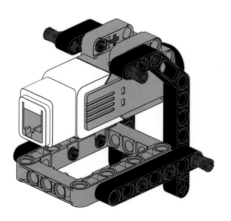

x2

10

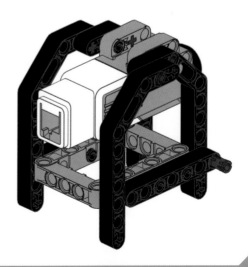

11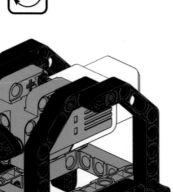

12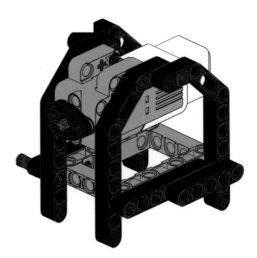

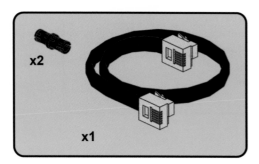

13

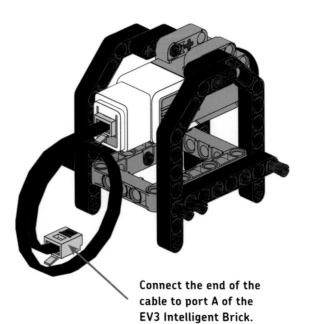

Connect the end of the cable to port A of the EV3 Intelligent Brick.

Make sure to position the remote's button underneath the cam so that the cam neither passes over the button nor gets hung up against the side of the remote. You can set this spacing in one of three ways:

* Raising or lowering the open center frame liftarm (placed in step 7) that serves as a platform for the remote control. The double bent liftarms on which the platform rides have several holes for setting the platform.

* Inserting plates under the remote control, in a process known as *shimming*, to raise it, as shown previously with a 4 × 6 plate in Figure 1-6.

* Using a different axle connection on the cam. The cam has four axle connections that can give the cam's spin different radii.

Spacing the camera and button properly will likely take some trial and error, especially if you use a different model or brand of camera and remote. It's easiest to adjust the button by manually turning the motor, rather than by turning on the power. To check that you've properly depressed the button, press it while the cam is engaged against the button. There shouldn't be any remaining downward travel or "click" left.

writing the software

You need software, or code, to control the Motion-Activated Critter Cam. The code will live in the EV3 Intelligent Brick, and you'll create it by using the LEGO MINDSTORMS EV3 environment, which you can download from *https://www.lego .com/en-us/themes/mindstorms/downloads/*.

After you install and open this program on your computer, you should see a colorful home page featuring the five basic robots of the MINDSTORMS EV3 #31313 set. If you hover your mouse over one of these robots, a menu should appear for instructions to build the example robot; building one or more of these robots is a good introduction if you're new to MINDSTORMS EV3. Instead of working with one of these robots, you'll be writing your own program from scratch, so go to the toolbar and select **File ▸ New Project**. You'll then find yourself at the starting point for the MINDSTORMS programming environment.

To write your program, you'll pick blocks representing specific functions from the drop-down menus and then chain these programming blocks together to create a flow of instructions. Figure 1-7 shows the final EV3 program.

The program will tell the EV3 Infrared Sensor to keep watching for the presence of an animal by searching for a proximity reading that indicates something is close to the sensor. When the sensor records such a proximity reading, the EV3 Medium Motor will spin to press the camera's remote control button. This watch-and-photograph routine will keep running until you shut off the program.

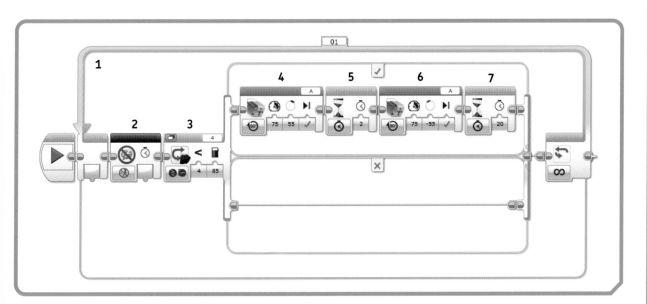

FIGURE 1-7: The program for the surveillance system consists of seven programming blocks.

The programming blocks perform the following functions:

1. **Loop block** This makes the watch-and-photograph routine repeat infinitely. Once the system detects and photographs an animal, the loop will reset to wait for the next animal to arrive.

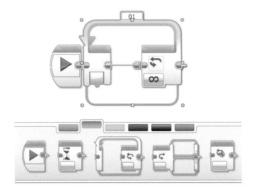

You can find the Loop block under the Flow Control tab at the bottom of the screen. Drag it next to the program start, and it should automatically connect to the program start symbol.

2. **Keep Awake block** This block keeps the EV3 Intelligent Brick from automatically shutting off, which happens in order to prevent the device's batteries from draining if it is accidentally left on. In this application, the EV3 Intelligent Brick should remain on constantly, waiting for an animal to arrive.

Drag a Keep Awake block into the loop. This block will automatically connect inside the loop when you bring the block into position. The Keep Awake block is under the Advanced tab at the bottom of the screen.

3. **Switch block** This provides the program with two possible sets of instructions by creating true and false paths. The instructions in the *true path* happen if a certain condition is true. The instructions in the *false path* happen if that condition is false. It's like flipping a switch. In this case, the condition is the proximity of an animal to the infrared sensor. If the sensor detects a proximity of greater than 85, the false path is triggered: nothing happens, and the loop starts over. If an animal gets close to the sensor, the sensor detects a proximity of less than 85, and the true path is triggered: the program executes blocks 4 through 7.

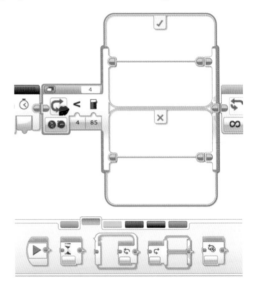

Join a Switch block to Keep Awake. From the Switch block's drop-down menus, set **Switch** to **Infrared Sensor-Compare-Proximity**, **Port** to **4**, **Compare Type** to **4**, and **Threshold Value** to **85**. You may need to adjust the threshold value for different setups, depending on the system's proximity to your intended target or brightness of the area.

4. **Medium Motor block** If the sensor detects something close, this block spins the motor connected to port A at 75 percent power clockwise for 55 degrees to push the camera remote's button. You might need different angles and powers for various models of cameras. For the Canon Rebel's remote control, 55 degrees gives a good, firm click of the remote's button.

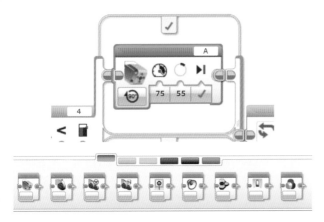

Place a Medium Motor block in the true branch of the Switch block. Set **Medium Motor** to **On for Degrees**, **Port** to **A**, **Power** to **75**, **Degrees** to **55**, and **Brake at End** to **True**.

5. **Wait block**   This block pauses for 2 seconds to make sure the motor has clicked the remote control button before resetting the motor to its previous position.

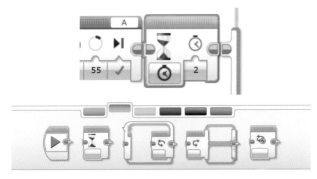

Connect a Wait block after the Medium Motor block. Set **Wait** to **Time**, and **Seconds** to **2**.

6. **Medium Motor block**   This block rotates the cam counterclockwise 55 degrees to set it back to its home position.

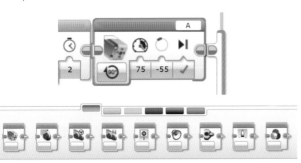

Connect a Medium Motor block after the Wait block. Set **Medium Motor** to **On for Degrees**, **Port** to **A**, **Power** to **75**, **Degrees** to **–55**, and **Brake at End** to **True**.

7. **Wait block**   This block pauses for 20 seconds before it allows the camera to take another picture. This waiting period keeps the Motion-Activated Critter Cam from repeatedly taking photos of the same animal as it lingers in the same area.

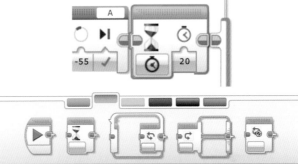

Add a Wait block to close out the true path of the Switch. Set **Wait** to **Time** and **Seconds** to **20**.

Once you've completed this program, download it into the EV3 Intelligent Brick's memory by connecting a USB cable from your computer to the EV3 Intelligent Brick, as shown in Figure 1-8. A USB cable comes with the MINDSTORMS EV3 #31313 set.

FIGURE 1-8: Use a USB cable to download programs from your computer onto the EV3 Intelligent Brick.

Be sure to have your EV3 Intelligent Brick turned on. It will take a few seconds for your computer to connect to the EV3 Intelligent Brick. Once it does, click the **Download** button in the lower-right corner of the MINDSTORMS EV3 programming environment (Figure 1-9).

FIGURE 1-9: Download your program onto the EV3 Intelligent Brick by clicking the Download button in the lower-right corner of the MINDSTORMS EV3 programming screen.

You should now be able to find the program in the EV3 Intelligent Brick's menu.

assembling the motion-activated critter cam

Now let's connect the infrared sensor head and button pusher to the EV3 Intelligent Brick. Press the infrared sensor head and button pusher into the sides of the EV3 Intelligent Brick, as shown in Figure 1-2. Use the connector cables to connect the EV3 Infrared Sensor to port 4 of the EV3 Intelligent Brick and the EV3 Medium Motor to port A of the EV3 Intelligent Brick. These port assignments match the ones set in the program.

With construction complete, you're ready to take the Motion-Activated Critter Cam outdoors.

using the motion-activated critter cam

Find out what types of birds visit your backyard by setting up your surveillance system at a feeder. Place the Motion-Activated Critter Cam on top of a ladder, with the EV3 Infrared Sensor aimed at one of the feed ports of a bird feeder. You may want to add to the height of the sensor stack for your particular setup—the building instructions have a two-brick-tall stack, but you may want to go higher. Then, point the remote control toward the camera on a tripod.

Make sure you've checked your camera for the following configuration options. First, your camera may have an automatic power-off feature that will turn off the camera if it sits idly for many minutes, to save battery life. You should deactivate this auto-off feature. Second, if you leave your camera set on autofocus, it may take a second or more to check the focus and may pulse the flash. This delay is too long to capture a bird, and the pulse flash can scare away animals. Set the focus once and then leave the camera on manual focus. This way, the camera takes a picture as soon an animal triggers the remote. Third, you may need flash when the sunlight level gets low. If you expect low light, leave the flash set to on and let the camera automatically determine the best exposure settings.

Figure 1-10 shows an example result.

FIGURE 1-10: A Carolina Chickadee is giving a stern look to a Downy Woodpecker.

You could also use your Motion-Activated Critter Cam to find out what nocturnal animals come to your backyard while you're asleep. To do this, turn on your camera's flash setting. Just as you did for the bird photos, set the camera focus once and then switch to manual focus mode for the duration of the experiment. Make sure to leave your camera out on a night with little chance of rain, dew, or frost in order to keep your equipment from being damaged. Also, the EV3 Intelligent Brick's status lights, which come on whenever a program is running, are rather bright in the dark of night. To keep these bright lights from scaring away animals, you could put a piece of opaque tape over the lights.

I used the Motion-Activated Critter Cam to figure out what kind of animal was scratching the post in Figure 1-1. I expected to find a raccoon or a possum, but I was wrong; after many nights of testing, neither a raccoon nor a possum showed up. But a cat visited on several occasions, as shown in Figure 1-11.

Case closed!

FIGURE 1-11: *A cat entered the fort shortly after midnight on an October night.*

ideas for going further

You could use the Motion-Activated Critter Cam to study other animals as well, such as hummingbirds, which are difficult to get close to and too small to see well from a distance. Setting up the Motion-Activated Critter Cam at a hummingbird feeder could capture great images of these elusive birds.

The invention in this chapter concentrated on photographing animals, but you could also build a device for taking actions other than triggering a camera. For example, the sensor could trigger an alarm to scare away the cat that is scratching the wood of the play fort in Figure 1-11. Or you could create a dynamic squirrel feeder that moves a cob of corn whenever a squirrel approaches. A motor could spin the corn out of the squirrel's reach, creating a puzzle for the squirrel to solve.

what you learned

In this chapter, you learned how to build a custom Motion-Activated Critter Cam that automatically records movement in a specific area. The system introduced you to EV3 sensors and motors, and you wrote your first program to control the system's behavior.

In the following chapters, you'll continue to use LEGO inventions for outdoor science studies.

2

the LEGO-compatible laser

Lasers fascinate people, and they're useful too; you can use them to project a beam over a long distance, make animated light shows, measure distances, experiment with the wave nature of light, and create holograms. It's natural to want to add a laser to LEGO inventions. Unfortunately, LEGO doesn't offer a LEGO-mountable laser, probably because of the safety issues discussed in the following box.

In this chapter, you'll build your own LEGO-mountable laser to use in future projects. This laser will use a LEGO-based battery source, so you can incorporate the power supply into your LEGO contraptions as well.

SAFETY WARNINGS

Keep in mind that lasers can harm your eyes. The laser used in this chapter is about as powerful as a laser pointer. While this laser beam won't necessarily damage people's eyes permanently, it can temporarily blind, distract, or disorient them, causing accidents, especially for drivers, athletes, pilots, or construction workers.

When working with lasers, stick to the following guidelines:

* Don't intentionally stare into the laser.

* Avoid shining the laser onto a mirror or reflective surface, since the beam could go in an unexpected direction.

* Don't let young kids play with the laser.

* Don't view the laser beam with a magnifying lens, binoculars, microscope, telescope, or other optical instruments. This can focus the intensity of the beam and potentially cause permanent vision loss.

* Attach a warning label to the laser so people are aware of the potential hazard. We typically classify lasers with the numbers 1 through 4, with 1 indicating the least hazardous and 4 indicating the most hazardous. At the 1 through 5 mW output power level used in this chapter, the LEGO-Compatible Laser will be a Class 2 or 3.

Drilling, cutting, and soldering can also cause injury. In this chapter, you'll be drilling holes into LEGO bricks, cutting objects with tools, and soldering parts, so make sure to use these tools safely by wearing safety glasses, securing LEGO bricks in a vise instead of holding them in your hand when drilling into them, and avoiding the very hot end of a soldering iron.

the diode laser module

To build the LEGO-Compatible Laser, you'll need a *diode laser module* (Figure 2-1), which includes a diode laser, along with a current-control circuit and a collimating lens.

Diode lasers are small devices, capable of producing bright beams, that are often used in CD and DVD players and laser pointers. The module's *current-control circuit* keeps the power output of the laser at a constant level. Otherwise, if the electrical power to the laser changed, the output of the laser could get too high, posing a safety hazard or burning out the laser, or too low, creating a beam too dim to be useful. This circuit also protects the laser in case you accidentally reverse the positive and negative connections. This particular laser module has control circuitry sticking out the back, but some devices keep the circuit inside the cylindrical enclosure. The *collimating lens* controls the divergence of the laser output to generate a narrow beam. Some laser modules have adjustable lenses to set a specific focus or divergence.

Some lasers, like the one shown in Figure 2-1, also come with stickers to place on the laser's future housing as a safety warning.

You can purchase diode laser modules at many places on the internet, including Digi-Key Electronics (*https://www.digikey.com/*), Newark (*https://www.newark.com/*), Jameco Electronics (*https://www.jameco.com/*), and Mouser Electronics (*https://www.mouser.com/*), for about $49. I recommend the Digi-Key VLM-635-02-LPA-ND, but the directions in this chapter are flexible for different devices.

Also keep in mind that jolts of static electricity can destroy diode lasers. If you find yourself picking up a lot of static in your home, touch something metal (for example, a doorknob) before working with the diode laser module. When you finish the module, it's a good idea to keep the completed module in the static control bag the laser comes in, so save this bag.

what you'll need

You'll need LEGO bricks to serve as a housing for your diode laser module. Figure 2-2 summarizes the bricks needed. None of these parts, except for 2780 Technic pins, are part of the MINDSTORMS EV3 #31313 set.

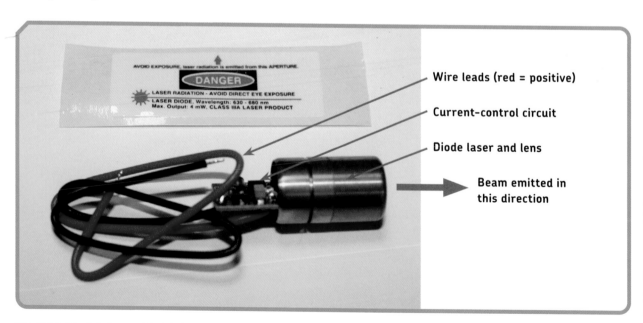

Wire leads (red = positive)

Current-control circuit

Diode laser and lens

Beam emitted in this direction

FIGURE 2-1: The diode laser module includes control circuitry, a collimating lens, and wire leads.

In addition to the LEGO bricks, you'll need the following components:

- Diode laser module (such as Digi-Key VLM-635-02-LPA-ND)

- Zener diode for the laser module (You can use 1N4733A for the diode laser module in this list, or for another laser module taking a 5V input.)

- 47 ohm, 1/2 watt resistor

- 9V battery

- Sugru moldable glue (one 5-gram single-use pack)

- Cyanoacrylate glue such as Super Glue, Krazy Glue, or Gorilla Glue (a few drops from a small tube)

You'll also need the following tools:

* Drill (I used a small cordless hand drill, though a drill press would be ideal if you already have one.)

* Drill bit (whose diameter is less than that of the diode laser module, usually about 6 mm or 1/4 inch)

* Tapered reamer of 3 mm (1/8 inch) to 12 mm (1/2 inch) capacity

* Soldering iron and solder

* Vise

* Diagonal cutters

* Wire strippers

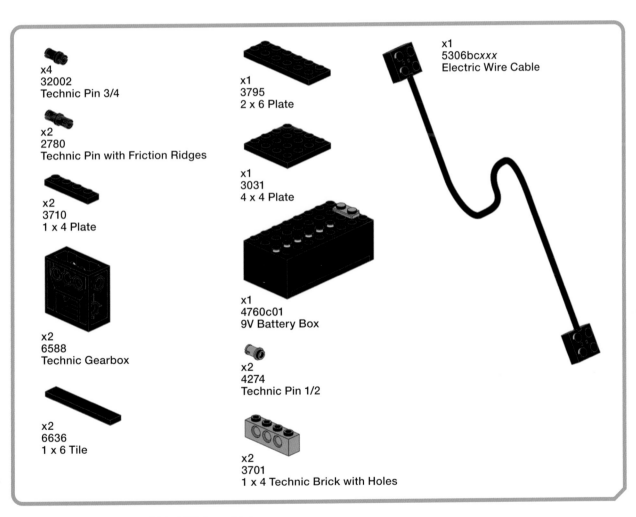

x4
32002
Technic Pin 3/4

x2
2780
Technic Pin with Friction Ridges

x2
3710
1 x 4 Plate

x2
6588
Technic Gearbox

x2
6636
1 x 6 Tile

x1
3795
2 x 6 Plate

x1
3031
4 x 4 Plate

x1
4760c01
9V Battery Box

x2
4274
Technic Pin 1/2

x2
3701
1 x 4 Technic Brick with Holes

x1
5306bc*xxx*
Electric Wire Cable

FIGURE 2-2: *LEGO parts used to build the LEGO-Compatible Laser*

mounting the laser module

To incorporate the diode laser module in LEGO creations, you need to mount it in a LEGO part. You'll use a Technic gearbox, shown in Figure 2-3, to hold it.

Gearboxes typically hold a worm gear against a conventional gear to drive the wheels of a vehicle, but the placement of its holes and its hollow lower portion are ideal for housing the diode laser module, as shown in Figure 2-4. Two gearboxes, held together with Technic pins, are used here to accommodate the full length of the diode laser module.

If you need more than two gearboxes, you can add a third one by using the process described next.

FIGURE 2-3: *A Technic gearbox*

FIGURE 2-4: *Two gearboxes joined together to house the length of the diode laser module*

drilling holes in the gearboxes

The gearboxes need modification to accommodate the laser module. Specifically, you need to drill out their uppermost center holes because their diameters are too small to fit the laser module. Drilling into plastic can be a little difficult, because the plastic tends to grab the drill bit, so it's a good idea to anticipate problems and buy several spare gearboxes. The drill bit for the job should be smaller than the diameter of the laser module. The diode laser module in Figure 2-1 has a diameter of 10 mm (2/5 inch), so I used an 8 mm (5/16 inch) drill bit.

The hole to enlarge with the drill is in the center, upper half of each gearbox. Use the existing LEGO-sized hole as a guide for the bit, as shown in Figure 2-5.

Since the laser module will go through both sides of each gearbox, drill through the hole on both sides. To keep both holes aligned, you should drill them from each side, rather than drilling on only one side and punching through the other end. Doing this would likely knock off the centering of each hole with respect to the other, because it's difficult to hold the drill straight and perpendicular through both holes.

If you have a drill press, you can use it to hold the bit straight and perpendicular to the gearbox hole. If you don't have one, a hand drill works fine, as long as you hold the drill perpendicular in a vise, as done in Figure 2-5. Don't try to hold the gearbox by hand when drilling, since it will spin out of control.

FIGURE 2-5: *Carefully hold the drill bit so that it is straight and perpendicular to the gearbox.*

rounding out the holes with a reamer

The holes you create will probably look messy and not quite round, with bits of plastic still clinging onto the rim. Use a reamer, shown in Figure 2-6, to clean up the holes and enlarge them to just the right size.

FIGURE 2-6: *After drilling the holes, widen them by twisting a reamer through them.*

Reamers come in different diameter ranges. The one to use here has an upper diameter of 12 mm (or 1/2 inch). Ream the hole a little at a time, stopping frequently to compare the hole's diameter to the diameter of the diode laser module.

Gradually open up the holes on both sides until the cylinder of the diode laser module fits snugly inside, as shown in Figure 2-7.

FIGURE 2-7: *The diode laser module should fit snugly inside the hole.*

The second gearbox (or third, if you need it) also needs to be drilled and reamed.

powering the laser module

To use your LEGO-Compatible Laser, you have to attach it to a power source. Most diode laser modules take input voltages of 3V to 6V. Unfortunately, the standard LEGO electric battery boxes have outputs of 9V, which isn't compatible with

the laser module. (LEGO made a 4.5V electric battery box in the 1960s and 1970s, but these old modules are hard to find in good condition, and they're quite large.)

To connect the battery box to the laser module, you'll modify the battery box with a few electrical components.

modifying an electric wire cable

The electric wire cables come in lengths ranging from 15 studs to 378 studs. The last few digits of the part number indicate the length in studs (where Figure 2-2 shows *xxx*). For example, 5306bc015 describes a wire cable that is 15 studs long. Choose a length based on how far you want your battery box to be from your laser. In Chapter 3, we'll conduct an experiment in which the diode laser module can be quite close to the battery box, so a short cable, such as the 5306bc036, should work fine.

Connect one end of the cable to the battery box. These cables contain two wires, a positive one and a negative one, side by side, so it's possible to keep track of which side of the cable is attached to which side of the battery box. In the orientation shown in Figure 2-8, the cable connector is attached so the wires are opposite the switch, and the right side of the box is negative and the left side is positive. That means the right-side wire is negative, and the left side is positive.

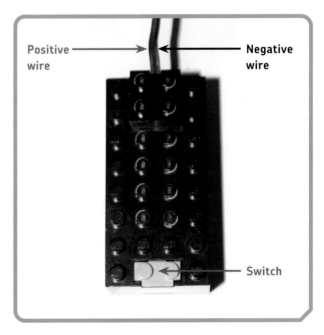

FIGURE 2-8: *A 9V battery box (4760c01) with an electric wire cable (5306bc) attached in this orientation has the positive connection on the left side and the negative connection on the right side.*

Next, cut off the LEGO connector on the other end of the cable—the end without the switch. This end has to be reduced to bare wires in order to be connected to the laser module. Clip off the LEGO connector on the laser end, as shown in Figure 2-9a. Then separate the two wires with diagonal cutters, as shown in Figure 2-9b. Using wire strippers, as shown in Figure 2-9c, expose the bare metal wire of both ends.

You can now thread the bare-wire end of the cable through a bottom, side hole of the rearmost gearbox, as shown in Figure 2-4. You'll attach it to the laser in a later step.

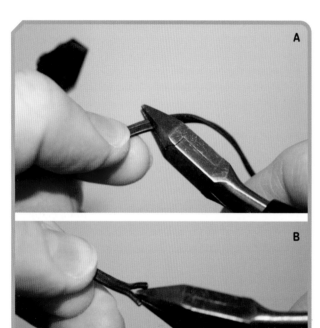

A

B

C

FIGURE 2-9: *Preparing the end of the wire cable for connection to the laser module*

connecting the Zener diode and resistor

In order to step down the 9V from the LEGO battery box so it's in the 3V to 6V range that the laser needs, you'll use a Zener diode and a current-limiting resistor (Figure 2-10).

FIGURE 2-10: *The Zener diode is the smaller device with a black band (left), while the resistor has multiple colored bands (right).*

A Zener diode creates a constant-output voltage from a higher-input voltage when reverse biased. *Reverse biasing* a diode means intentionally hooking it up backward. Using the Zener diode to connect the battery pack to the laser in this way will put the voltage in the proper range. Zener diodes come in different voltage outputs, so select a Zener diode that matches the voltage of the diode laser. For example, the IN4733A Zener diode has an output of 5.1V that fits within the voltage range that the laser diode module needs.

But if you connected the Zener diode to the power source by itself, it would consume all the electrical current the source could produce and likely burn out. Just as you have to limit the amount of food given to a pet goldfish, you have to limit the current so that the Zener diode won't eat its way to self-destruction. To do this, you'll place a resistor in front of the diode. Be sure to get a resistor with the 1/2 watt power rating, as opposed to the also-common 1/4 watt variety. A 1/4 watt resistor used here could possibly melt, causing the entire circuit to fail.

The circuit to put the diode and the resistor together is drawn in Figure 2-11.

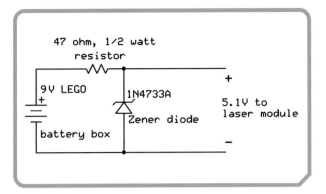

FIGURE 2-11: Use a Zener diode to step down the 9V from the LEGO battery box to a suitable voltage for the laser diode module.

To implement this circuit, you have to do a little soldering. If you don't know how to solder, video tutorials on YouTube (*https://www.youtube.com/*) and SparkFun Electronics (*https://www.sparkfun.com/*) can help. You can buy the soldering iron and accessories at SparkFun and Jameco. You need to make three solder joints, pictured in Figure 2-12, and then trim the leftover lengths of the wire leads.

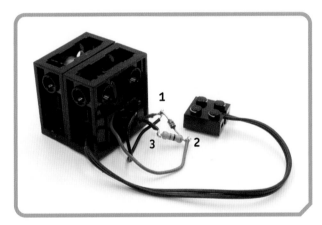

FIGURE 2-12: Three solder joints connect power to the laser diode module.

Before soldering, make sure that the wire cable from the battery box passes through a bottom, side hole of the gearbox and that the two wires from the laser module travel down through the bottom, back of the gearbox. Later, you'll tuck the soldered joints and Zener diode circuit into the bottom of the gearbox.

The three solder joints you need to make are as follows:

Solder joint 1 Three leads consisting of the negative lead from the LEGO battery box, the black lead of the laser module, and the anode end of the Zener diode. The anode end of the Zener diode is the one *without* the black band. *Anode* is a term to describe the electrically positive end of an electronic component. But remember that a Zener diode is intentionally hooked up backward, so the anode of the Zener diode goes to the negative end of the battery.

Solder joint 2 Three leads consisting of one end of the resistor, the red lead of the laser module, and the cathode end of the Zener diode. The cathode end of the Zener diode has a black band around it. The *cathode* is the opposite end from the anode, corresponding to the negative end of an electronic component, except in the circuit used here with the Zener diode being used in a reverse bias. Resistors have no polarity, so it doesn't matter which end of the resistor you use for this step.

Solder joint 3 Two leads, consisting of the end of the positive lead from the LEGO battery box and the other end of the resistor.

After you've trimmed away the excess lead wires to look like Figure 2-12, it's time to light up the laser. Put a battery into the LEGO battery box and turn on the switch to the battery box. Your reward should be a nice bright beam. If your laser doesn't turn on, check that your circuit is connected correctly to match the circuit diagram of Figure 2-11, paying special attention to the polarity of the Zener diode—the end of the Zener diode with the black band around it should connect to the red, positive lead of the laser module.

securing the laser and wiring

When you turn on the laser, you should see that the output beam has an elliptical shape. Rotate the laser module until you have an orientation that you like. You may have to partially disassemble the two gearboxes' halves to do this.

Once the elliptical beam is oriented to your liking and the gearboxes are pressed back together, glue the diode laser module into place to keep it from shifting out of position. Super Glue (or other cyanoacrylate glue such as Krazy Glue or Gorilla Glue) holds tight, so use only a small drop in case you need to remove the glue later. Figure 2-13 shows the placement of a drop of Super Glue around the rim of the front of the diode laser module. Use three such drops around the rim.

FIGURE 2-13: *Apply a few drops of Super Glue to keep the laser from moving.*

Once the Super Glue dries, you can tend to the electrical connections. In order to protect the bare wires, keep them from shorting out, and follow a good safety practice of not leaving live electrical circuits exposed, cover all the electronics with Sugru moldable glue in a technique known as *encapsulation*. Sugru feels like modeling clay when fresh from the package, but it hardens into a rubbery form. With the circuit turned off, wrap Sugru around the electronics, as shown in Figure 2-14.

FIGURE 2-14: *Cover all the electronic components and bare wires with Sugru.*

When fully encapsulated, the Sugru blob should have a lima bean shape, as in Figure 2-14, and you should be able to tuck it away inside the hollow lower section of the gearbox. Pack the remaining Sugru around the hole of the gearbox where the LEGO wire cable enters, as shown in Figure 2-4. This relieves strain on the wire cable by keeping the wires from rubbing around or twisting.

adding a cover to the laser structure

With the electronics encapsulated and tucked into the gearbox, you can now add a cover for the laser module housed within the gearboxes. This cover, shown in Figure 2-15, will hide and protect the Zener diode circuit and give strength to the LEGO-Compatible Laser as you attach it to various LEGO inventions.

FIGURE 2-15: *Done! Your LEGO-Compatible Laser is complete and ready to use.*

The cover also provides a place for the warning label sticker pointing at the laser output. Building instructions for the cover are shown here.

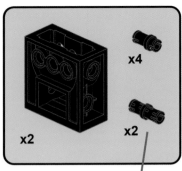

1

Pins, placed when you installed the laser module, hold the gearboxes together.

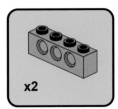

2

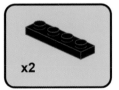

3

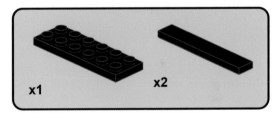

4

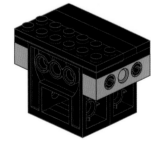

x2

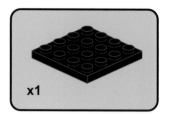

x1

5

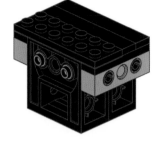

6

what you learned

In this chapter, you built a laser that is compatible with LEGO constructions. You learned how to work with diode laser modules, by building a LEGO housing for the laser and an electrical circuit to provide power from a LEGO battery box. You can now put the LEGO-Compatible Laser to use, starting in the next chapter with a laser-based security system.

3

the laser security fence

In this chapter, you'll build an intrusion detection system that sounds an alarm if an intruder crosses a certain boundary. For it to be useful, you'll need to build a device that can detect objects at a distance of several meters—the width of the approach to a doorway, at least. The MINDSTORMS EV3 Infrared Sensor, used in Chapter 1 for the Motion-Activated Critter Cam, can detect only objects that come pretty close to the sensor. To warn you of approaching people, you'll build a Laser Security Fence that uses the LEGO-Compatible Laser from the previous chapter to project a line and then sounds an alarm if an intruder crosses that line.

building the laser security fence

The Laser Security Fence is composed of two parts: a laser that projects a laser beam and a sensor to detect changes in the beam's intensity. For the laser, you'll use the LEGO-Compatible Laser you built in Chapter 2. For the sensor, you'll build a MINDSTORMS EV3 sensing device. Figure 3-1 shows both of these parts.

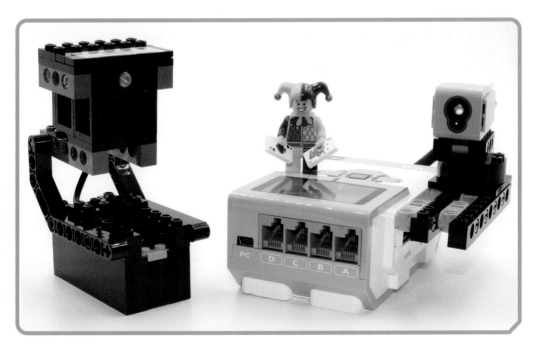

FIGURE 3-1: The Laser Security Fence uses a Color Sensor to detect interruptions of a laser beam.

The sensing device will use a Color Sensor, part of the MINDSTORMS EV3 #31313 set, attached to the side of the EV3 Intelligent Brick to detect the light beam from the laser. If intruders cross the fence, they'll interrupt the beam, triggering an alarm and flashing lights.

what you'll need

You'll use the LEGO parts in Figure 3-2 to create a structure to hold and tilt the laser. This figure also includes the LEGO components to hold the Color Sensor. An asterisk indicates a LEGO part not found in the MINDSTORMS EV3 #31313 set.

You'll also need the LEGO-Compatible Laser built in Chapter 2, including its battery box.

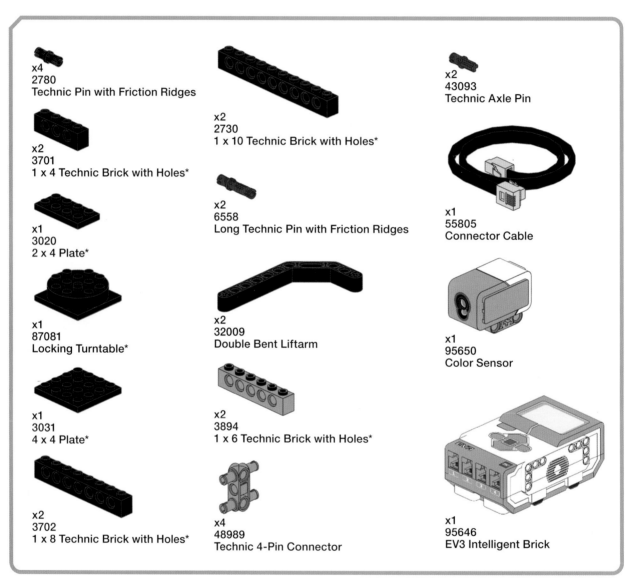

x4
2780
Technic Pin with Friction Ridges

x2
3701
1 x 4 Technic Brick with Holes*

x1
3020
2 x 4 Plate*

x1
87081
Locking Turntable*

x1
3031
4 x 4 Plate*

x2
3702
1 x 8 Technic Brick with Holes*

x2
2730
1 x 10 Technic Brick with Holes*

x2
6558
Long Technic Pin with Friction Ridges

x2
32009
Double Bent Liftarm

x2
3894
1 x 6 Technic Brick with Holes*

x4
48989
Technic 4-Pin Connector

x2
43093
Technic Axle Pin

x1
55805
Connector Cable

x1
95650
Color Sensor

x1
95646
EV3 Intelligent Brick

FIGURE 3-2: Parts used to build the Laser Security Fence

building the LEGO-compatible laser mount

The LEGO-Compatible Laser mount, shown on the left side of Figure 3-1, takes advantage of the weight of the battery box to create a solid base for the laser. Follow these steps to build it.

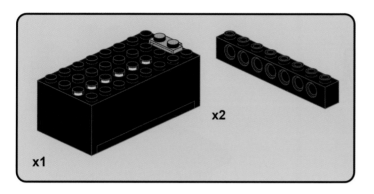

x1

x2

1

x4

2

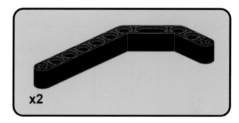

x2

3

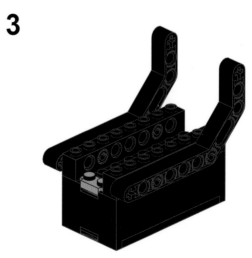

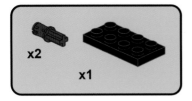

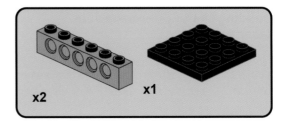

4

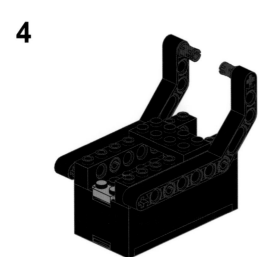

5

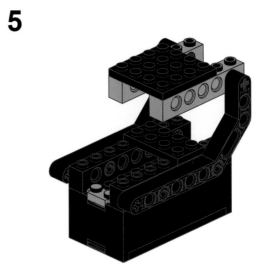

6

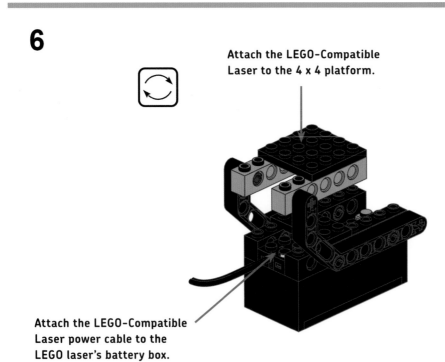

Attach the LEGO-Compatible
Laser to the 4 x 4 platform.

Attach the LEGO-Compatible
Laser power cable to the
LEGO laser's battery box.

building the sensor mount

Use the following steps to build a mount for the EV3 Color Sensor. You'll attach this mount, in turn, to the side of the EV3 Intelligent Brick, as shown in Figure 3-1. The Color Sensor's mount features a locking turntable that helps align the sensor to the laser beam.

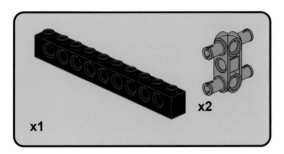

x1   x2

1

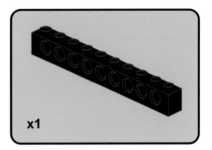

x1

2

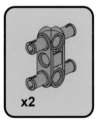

x2

3

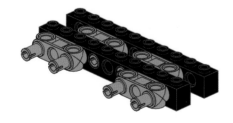

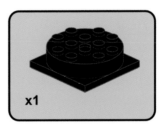

x1

4

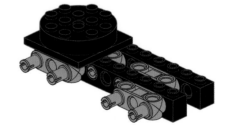

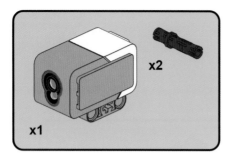

5

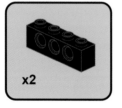

6

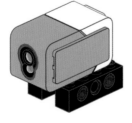

7

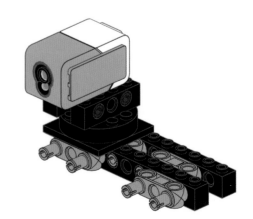

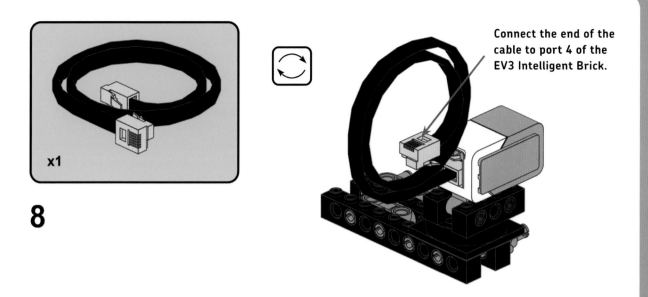

Connect the end of the cable to port 4 of the EV3 Intelligent Brick.

8

writing the software

You'll control the Laser Security Fence by programming the EV3 Intelligent Brick. This program, shown in Figure 3-3, will monitor the light-level reading from the Color Sensor to determine whether the laser beam drops in intensity, which should occur if something crosses the beam.

The Color Sensor has many modes of operation for measuring or comparing color, reflected light intensity, and ambient light intensity. In conjunction with a Switch block, you'll use a mode of operation called *Color Sensor–Compare–Ambient Light Intensity*, which measures the brightness of the light on the face of the Color Sensor.

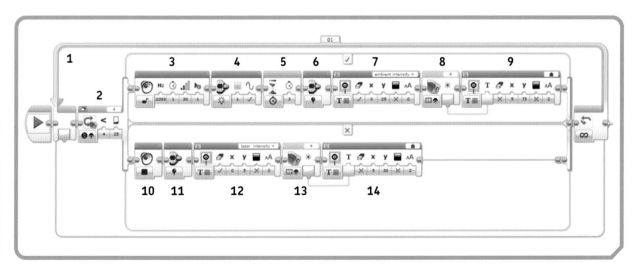

FIGURE 3-3: The program for the Laser Security Fence is based on the Color Sensor–Compare–Ambient Light Intensity mode of a Switch block.

If you're new to programming MINDSTORMS EV3, you can find instructions to get started in Chapter 1. This program uses the following blocks:

1. **Loop block**   This makes the program repeat infinitely. This means the program will constantly test for whether the laser beam is striking the Color Sensor.

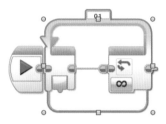

Drag a Loop block onto the screen and place it next to the Start block in the programming area.

2. **Switch block**   This block creates true and false paths based on readings from the Color Sensor. You'll set this block to Color Sensor–Compare–Ambient Light Intensity. In this setting, the Switch block tests whether the light intensity seen by the Color Sensor is above or below a threshold level. If the light intensity level seen by the Color Sensor is less than 25, the true path of blocks 3 through 9 is taken. If the level is 25 or higher, the false path of blocks 10 through 14 is taken.

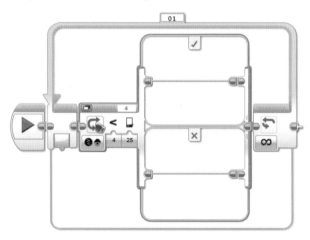

Place the Switch block inside the Loop block. Set **Switch–Color Sensor** to **Color Sensor–Compare–Ambient Light Intensity**, **Port** to **4**, **Compare Type** to **4**, and **Threshold Value** to **25**. You may need to adjust the threshold level depending on the background light level (whether you're inside a dark room or

outdoors in the sun) or the laser you're using. In bright sunlight, you should set the threshold to about 70. For help determining how to set the threshold, check the ambient light reading on the Display block in step 9. The threshold should always be higher than that value.

3. **Sound block**   Place this block in the true path created by the Switch block. If the Switch block reports the true condition because the laser beam is interrupted, this block begins playing a sound on the EV3 Intelligent Brick's speaker. You can select various beeps or musical tones from the Sound block's drop-down menus.

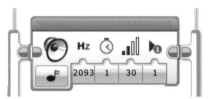

Place a Sound block in the true path of the Switch block. Set **Sound** to **Play Tone**, **Frequency** to **2093**, **Duration** to **1**, **Volume** to **30**, and **Play Type** to **1**.

4. **Brick Status Light block**   After the Sound block plays, this block flashes a yellow light on the front of the EV3 Intelligent Brick to indicate an alarm condition. If you prefer, you can make this light red or green.

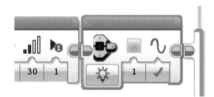

Place the Brick Status Light block after the Sound block. Set **Brick Status Light** to **On**, **Color** to **1**, and **Pulse** to **True**.

5. **Wait block**   This block keeps the yellow alarm light on for 3 seconds, giving you time to notice that an alarm event has occurred.

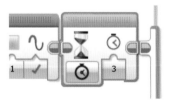

Connect a Wait block after the Brick Status Light block. Set **Wait** to **Time** and **Seconds** to **3**.

6. **Brick Status Light block**   This block turns off the yellow alarm light at the end of the alarm warning event.

Connect the Brick Status Light block after the Wait block. Set **Brick Status Light** to **Off**.

7. **Display block**   To help set the threshold value in the Switch block, the Display block creates a display on the EV3 Intelligent Brick's front panel that shows the current ambient light level. You'll add text to this block to display the message *ambient intensity*. This will serve as a label for the light level detected by the sensor in the next step.

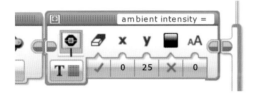

Connect the Display block after the Brick Status Light block. Set **Display** to **Text–Pixels**, **Text** to **ambient intensity =**, **Clear Screen** to **True**, **x** to **0**, **y** to **25**, **Color** to **False**, and **Font** to **0**. The x- and y-coordinates in these menus determine the position where the text will appear on the front panel display of the EV3 Intelligent Brick.

8. **Color Sensor block**   This measures the intensity of the light detected by the sensor when the laser light has been blocked. This measurement becomes useful when you decide which threshold value to set in block 2. The threshold should be a little above the background light level measured in this step. You'll display the measurement in the following step.

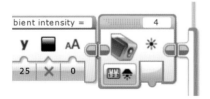

Attach the Color Sensor block after the Display block. Set **Color Sensor** to **Measure–Ambient Intensity** and **Port** to **4**.

9. **Display block**   This block shows the ambient light intensity measured by the Color Sensor. It uses a wire connection to take the measured light intensity from block 8 as a text input into the Display block. The measured value should appear on the EV3 Intelligent Brick's front panel, to the right of the text label created in block 7.

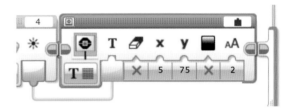

Place the Display block after the Color Sensor block. Set **Display** to **Text–Pixel**, **Text Type** to **Wired**, **Clear Screen** to **False**, **x** to **5**, **y** to **75**, **Color** to **False**, and **Font Size** to **2**. Connect a wire between the output of the Color Sensor block and the input of the Display block. To make the wire connection, hover your mouse over the output of the Color Sensor. An icon of a spool of wire should appear. Click and hold the Color Sensor output and drag your cursor to the input of the Display block. The wire connection should then appear as a yellow line.

10. **Sound block**   In the false path, which occurs as long as the laser beam is hitting the Color Sensor, turn off all alarm sounds by inserting this Sound block. This block ensures that the sound from the alarm doesn't continue after a previous alarm occurrence.

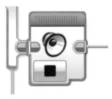

Place the Sound block into the false path of the Switch block and set **Sound** to **Stop**.

11. **Brick Status Light block**   This block turns off the indicator light on the EV3 Intelligent Brick front panel, because there is no alarm condition in the false path.

Place a Brick Status Light block after the Sound block and set **Brick Status Light** to **Off**.

12. **Display block**   Similar to what you programmed in the true path, this block creates a text label that displays the message *laser intensity* to identify the reading from the Color Sensor.

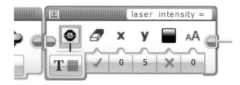

Place the Display block after the Brick Status Light block. Set **Display** to **Text–Pixels**, **Text** to **laser intensity =**, **Clear Screen** to **True**, **x** to **0**, **y** to **5**, **Color** to **False**, and **Font** to **0**.

13. **Color Sensor block**   This measures the intensity of the laser light seen by the sensor when the laser is illuminating the Color Sensor. The value measured here can help you align the sensor to the laser beam, as you'll see in the next section.

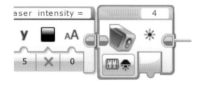

Attach the Color Sensor block after the Display block. Set **Color Sensor** to **Measure–Ambient Intensity** and **Port** to **4**. The output from the Color Sensor will be fed to the next block in the program sequence.

14. **Display block**   This block shows the laser light intensity measured by the Color Sensor block, using the measurements from the Color Sensor as input to display on the EV3 Intelligent Brick.

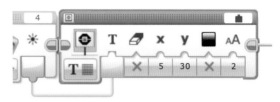

Place the Display block after the Color Sensor block. Set **Display** to **Text–Pixels**, **Text Type** to **Wired**, **Clear Screen** to **False**, **x** to **5**, **y** to **30**, **Color** to **False**, and **Font Size** to **2**. Connect a wire between the output of the Color Sensor block and the input of the Display block.

After completing the program, download it into the memory of the EV3 Intelligent Brick. If you're not familiar with the download process, review it in Chapter 1. Your Laser Security Fence is now ready for installation.

using the laser security fence

Place the LEGO-Compatible Laser and Color Sensor on structures of the same height so the laser beam aligns with the face of the Color Sensor. You may need to tilt the laser by a pivoting the joint built into the mount.

Figure 3-4 shows the Laser Security Fence set up across two exterior doors and a window at the back of a house.

In this example, the two ends of the Laser Security Fence sit on handrails, and the laser covers a distance of 12 meters. An attempted entry at either of the two doors or the window sets off the alarm. Notice that the mark of the laser beam on the face of the Color Sensor is larger than the beam when it left the laser. All laser beams expand over distance, a property called *divergence*.

With setups that cover 10 meters or more, you might struggle to align the laser beam onto the Color Sensor by tipping the laser mount. Try adjusting the angle of the laser and then moving the light sensor into the beam path. You can also place LEGO bricks under the EV3 Intelligent Brick to raise the height of the light sensor, as shown in Figure 3-4.

FIGURE 3-4: *The Laser Security Fence guards the back of a house.*

If you plan to install the Laser Security Fence outside for a long time, you could make LEGO brick housings to protect the laser source and EV3 Intelligent Brick. Alternatively, you could place one or both ends of the setup indoors with a view through a window, as shown in Figure 3-5.

FIGURE 3-5: *You can place one or both ends of the Laser Security Fence indoors with a view through a window.*

In this configuration, the sensor is indoors, and the laser is sitting on top of a fence post outside. Transmitting the laser beam through the window glass weakens the beam's intensity, so it hits the Color Sensor at about 90 percent of its normal value. Most of the loss comes from the reflection of the laser beam off the glass. You may need to adjust the threshold level in the Switch block of the program to account for the lost laser intensity.

ideas for going further

Instead of triggering an alarm, the Laser Security Fence could activate another protective action. You could change the code so it activates a camera to record video or pictures, as you did in Chapter 1. You could also turn on security lighting by using the IoT Control Relay device described in Chapter 6. Instead of using the laser beam in a straight line, the Laser Security Fence could protect the area around a corner by using a mirror. You'll learn about using a mirror with a laser beam in Chapter 11.

The laser used in this chapter has a visible red color, so it's possible for an intruder to see the beam. If you want to avoid detection, you could use an invisible infrared laser. The Color Sensor end of the Laser Security Fence will still work with an infrared laser as long as its wavelength is below 1,100 nanometers. Aligning an infrared beam requires special tools and techniques, and Chapter 5 explains how to work with them.

what you learned

In this chapter, you used a laser to build the Laser Security Fence, which can detect intrusion or motion over a long distance—much longer than a MINDSTORMS EV3 Infrared or Ultrasonic Sensors can. You also learned to use a MIND-STORMS Color Sensor to detect various light levels.

4

Morse code transmitters

In the 19th century, people used *Morse code* to transmit messages by telegraph, a device that creates an audible tone in a receiver. To send a message in Morse code, you encode letters and numbers by using a combination of short tones (called *dots*) and long tones (called *dashes*). Numbers are composed of five tones, and letters are composed of no more than four tones. Figure 4-1 lists the codes for every letter and number.

While telegraphy has long passed from daily communications, people still use Morse code in certain emergencies or on occasions when sophisticated communication technology fails. To send a Morse code message, a person can blink a flashlight, raise flags in either hand, or tap on pipes. Famously, Jeremiah Denton, a captured US Navy officer, revealed that he was being tortured by blinking his eyes in Morse code during a forced television interview. You might want to learn Morse code for your own emergency preparedness.

This chapter will help you become proficient at Morse code by showing you how to build a Manual Message Transmitter, which uses lights and tones to transmit a Morse code message that you tap out on a sensor. You'll then build an Automated Message Transmitter that you can program to transmit the same Morse code message in a loop. In both cases, you'll be able to read the messages by recording the dots and dashes displayed on the EV3 Intelligent Brick's front panel lights or played on its built-in speaker.

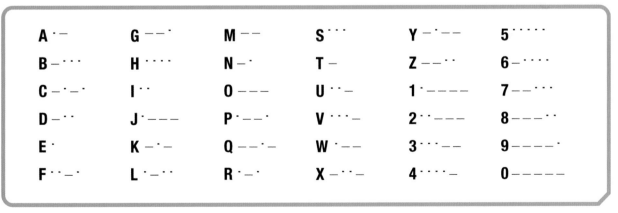

FIGURE 4-1: Morse code characters

what you'll need

Figure 4-2 summarizes the parts needed for the Morse code projects in this chapter.

You can find all the parts in the MINDSTORMS EV3 #31313 set.

building a manual message transmitter

A good way to build Morse code proficiency is to practice keying in letters and messages. The Manual Message Transmitter you'll build in this section, shown in Figure 4-3, uses a Touch Sensor, as well as the EV3 Intelligent Brick's speaker and lights.

You'll hold the Touch Sensor down briefly to produce a dot and at length to produce a dash. The EV3 Intelligent Brick should make a corresponding sound and light up its status lights.

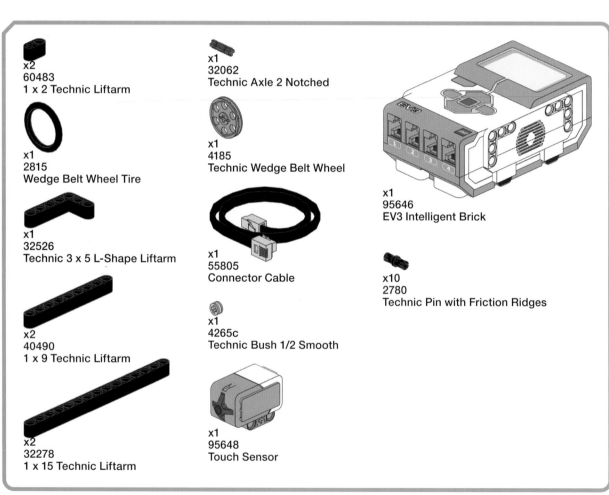

x2
60483
1 x 2 Technic Liftarm

x1
2815
Wedge Belt Wheel Tire

x1
32526
Technic 3 x 5 L-Shape Liftarm

x2
40490
1 x 9 Technic Liftarm

x2
32278
1 x 15 Technic Liftarm

x1
32062
Technic Axle 2 Notched

x1
4185
Technic Wedge Belt Wheel

x1
55805
Connector Cable

x1
4265c
Technic Bush 1/2 Smooth

x1
95648
Touch Sensor

x1
95646
EV3 Intelligent Brick

x10
2780
Technic Pin with Friction Ridges

FIGURE 4-2: Parts used to build the Morse code projects

FIGURE 4-3: *The Manual Message Transmitter is built from a Touch Sensor attached to the side of the EV3 Intelligent Brick.*

attaching the sensor and Intelligent Brick

To make the Manual Message Transmitter, you'll attach the Touch Sensor to the side of the EV3 Intelligent Brick with an L-shaped liftarm. You'll add a wheel on top of the sensor to press down. Use the following instructions to attach the Touch Sensor.

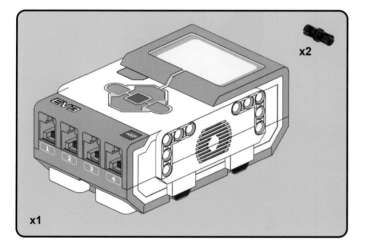

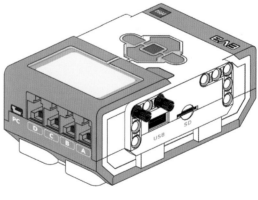

1

2

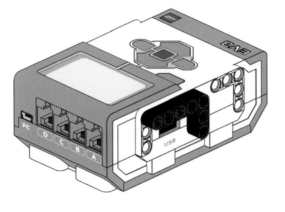

3

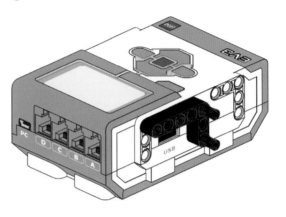

4

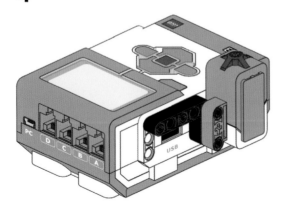

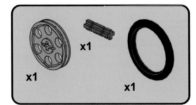

5

6

7

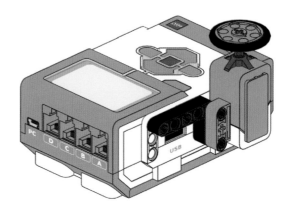

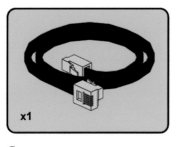

8

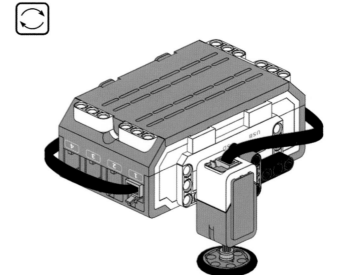

writing the manual message transmitter software

You'll need to program the EV3 Intelligent Brick to translate presses of the Touch Sensor into Morse code signals. The software should continuously monitor whether the Touch Sensor is being pressed. When someone presses the sensor, a light should go on and a tone should play. The program, composed of five blocks, is shown in Figure 4-4.

To build the program, use the following blocks:

1. **Loop block**   This infinitely repeats the block that checks for input from the Touch Sensor.

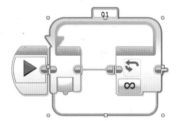

Click the Loop block in the Flow Control tab at the bottom of the screen and drag it to the right of the Start block.

2. **Switch block**   This creates true and false paths based on whether someone presses the Touch Sensor. If someone presses the Touch Sensor, the true path runs. Otherwise, the false path runs.

FIGURE 4-4: The Manual Message Transmitter is based on a Touch Sensor Switch block.

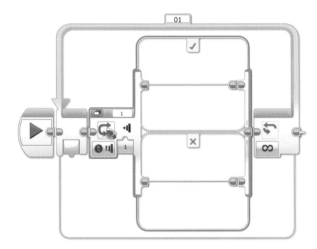

Place the Switch block inside the Loop and set **Switch-Touch Sensor** to **Touch Sensor-Compare-State**, **Port** to **1**, and **State** to **1**.

3. **Brick Status Light block**   This turns on the light on the front panel of the EV3 Intelligent Brick. This light will stay on as long as the Touch Sensor is pressed.

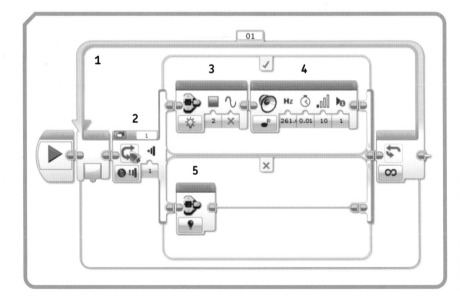

Select the Brick Status Light block from the Action tab and place it inside the true path of the Switch block. Set **Brick Status Light** to **On**, **Color** to **2**, and **Pulse** to **False**.

4. **Sound block**   This emits a tone from the EV3 Intelligent Brick's speaker for as long as the Touch Sensor is pressed.

Place the Sound block after the Brick Status Light block and select the drop-down options for **Sound** to **Play Tone**, **Frequency** to **261.626**, **Duration** to **0.01**, **Volume** to **10**, and **Play Type** to **1**.

5. **Brick Status Light block**   This begins the false path of the Switch block by turning off the light on the EV3 Intelligent Brick front panel.

Place a Brick Status Light block into the false branch of the Switch block, setting its drop-down menu option for **Brick Status Light** to **Off**.

using the manual message transmitter

Learning Morse code by keying in letters by hand will help you memorize the characters much more easily than just studying a list. Practice keying in letters on the Touch Sensor until you get the hang of it. Then use it to type messages to friends. You may notice that when you press the Touch Sensor, the Manual Message Transmitter generates a musical static rather than a clean tone. This altered tone occurs because the EV3 Intelligent Brick is set up to play only one sound at a time, whereas you're continuously playing a sound. This shouldn't affect your Morse code messages.

building an automated message transmitter

You may sometimes want your Morse code message to repeat continuously. For example, you might want to send a message to someone who walks by at a random time. The Automated Message Transmitter lets you write a Morse code message as part of the software, and then repeats that message forever.

This project uses the EV3 Intelligent Brick and a stand to keep the EV3 Intelligent Brick from falling over. The EV3 Intelligent Brick's status lights will blink out the message in green (Figure 4-5). They'll also blink a red light to signify the end of a letter and a yellow light to signify the beginning of the message.

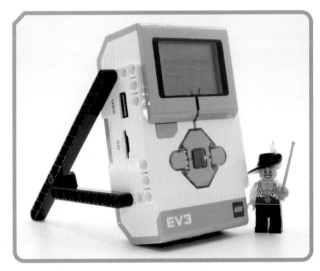

FIGURE 4-5: The Automated Message Transmitter blinks out messages on the EV3 Intelligent Brick's status lights.

Use the following building instructions to build the stand.

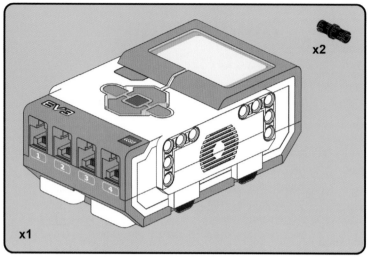

1

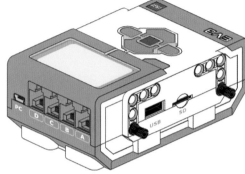

2

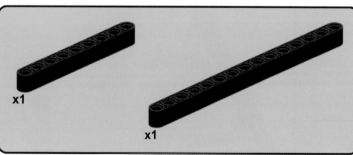

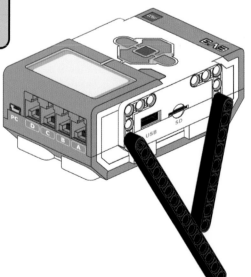

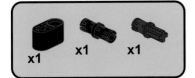

3

4

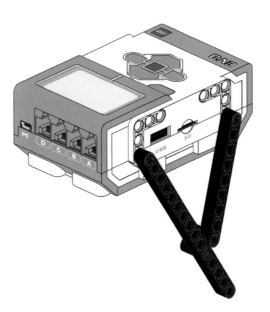

5

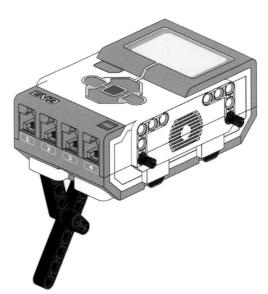

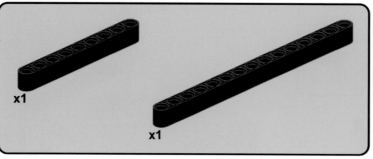

6

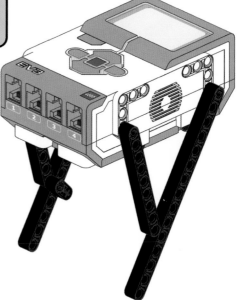

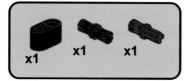

7

8

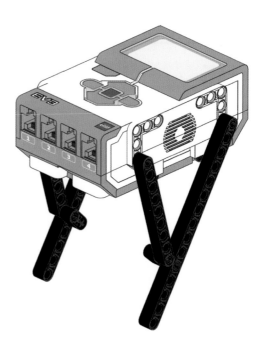

writing the dot and dash programs

Instead of sounding a tone and flashing a light when it receives outside input, the Automated Message Transmitter must generate signals of the appropriate length based on a message you write as part of the code. That means you'll have to create two programs to represent a dot and dash and then collect each of these programs within a My Block, a programming feature in the MINDSTORMS EV3 environment that lets you collect the blocks of a program into a single block.

Using My Blocks keeps you from having to repeat long sequences of blocks every time you want to reuse sections of code. In this project, you'll be making My Blocks for a dot and dash, and then using the dot and dash My Blocks to create a My Block for each letter of the alphabet. Finally, you can write your message by stringing together letter My Blocks.

Let's start by writing the dot and dash programs, shown in Figure 4-6. The two programs are identical except for the length of time that the status light stays on.

The program uses the following blocks:

1. **Brick Status Light block**   This turns on the EV3 Intelligent Brick status light in the color green.

Select a Brick Status Light block from the Action tab at the bottom of the screen and place it after the program start. Set **Brick Status Light** to **On**, **Color** to **0**, and **Pulse** to **False**.

2. **Wait block**   This keeps the green status light on for 0.2 seconds for a dot and 0.75 seconds for a dash. The dot version is shown here.

Select a Wait block from the Flow Control tab and place it after the Brick Status Light block. Set **Wait** to **Time** and **Seconds** to **0.2** for a dot or **0.75** for a dash.

3. **Brick Status Light block**   This turns off the green status light.

Attach a Brick Status Light block after the Wait block and set **Brick Status Light** to **Off**.

4. **Wait block**   This creates a pause of 0.5 seconds before moving on to the next dot or dash.

Place the Wait block after the Brick Status Light block. Set **Wait** to **Time** and **Seconds** to **0.5**.

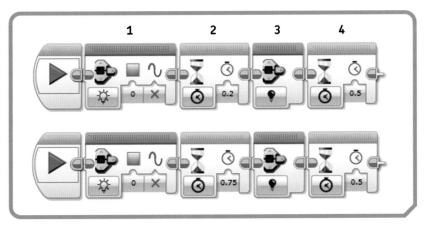

FIGURE 4-6: The upper block set creates a dot, and the lower block set creates a dash.

creating dot and dash blocks with My Blocks

Once you've completed the dot and dash programs, make a My Block of each program. First, left-click the first block in each program. Then, hold the SHIFT key and click the other three blocks to highlight the entire program. Don't include the Start block that begins the program. With all blocks of the program highlighted, go to the toolbar at the top of the screen and select **Tools ▸ My Block Builder**. The My Block Builder menu should appear on your screen, as shown in Figure 4-7.

Your custom block will perform the same function as the program you highlighted. My Block Builder allows you to assign a name, description, and icon to your block. You can load a personalized drawing to use as the My Block's icon or choose one of the icons already provided. Make sure you'll be able to distinguish the dot block from the dash block.

When you've entered your My Block information, click **Finish** on My Block Builder. Your new My Block will appear under the My Blocks tab at the bottom of the screen. You can use these My Blocks as you would any other programming block in the MINDSTORMS EV3 programming environment.

creating letter blocks

Use the dot and dash blocks to create programs for every number and letter in the alphabet. Figure 4-8 shows programs for the letters A and B.

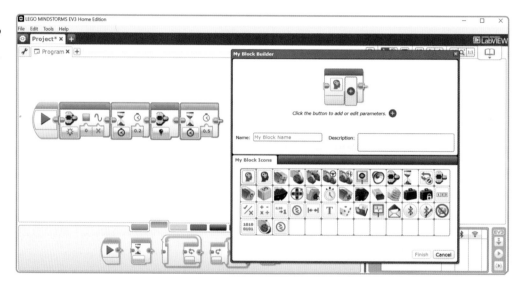

FIGURE 4-7: *The My Block Builder menu allows you to assign names, icons, and properties to your custom-made block.*

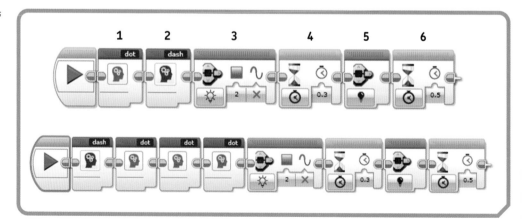

FIGURE 4-8: *The programs for the letters A and B*

To create the A program, use the following six blocks:

1. **Dot block**   This blinks a dot. (Remember that A in Morse code is represented by a dot and a dash.)

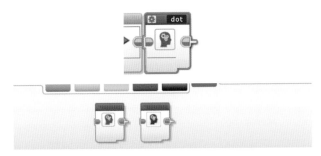

Select the dot block from the My Blocks tab at the bottom of the screen and drag it next to the Start block.

2. **Dash block**   This blinks a dash to complete the letter A.

Add the dash block after the dot block.

Blocks 3 through 5 blink a red light to indicate the end of a letter, which should help viewers decode the message:

3. **Brick Status Light block**   This turns on the red status light on the front of the EV3 Intelligent Brick.

Place the Brick Status Light block after the dash block. Set **Brick Status Light** to **On**, **Color** to **2**, and **Pulse** to **False**.

4. **Wait block**   This pauses for 0.3 seconds to keep the red light on.

Place a Wait block after the Brick Status Light block. Set **Wait** to **Time** and **Seconds** to **0.3**.

5. **Brick Status Light block**   This turns off the red end-of-letter indicator after the Brick Status Light has been on for 0.3 seconds.

Join a Brick Status Light block to the Wait block and set **Brick Status Light** to **Off**.

6. **Wait block**   This creates a pause of 0.5 seconds before the next letter. This pause gives the message reader time to get ready to interpret the next letter.

Conclude the program by adding a Wait block and setting **Wait** to **Time** and **Seconds** to **0.5**.

Follow this procedure to write a program for every letter and number. Replace the dot and dash of the A block with the combination of dots and dashes that forms each letter and number. Then turn each program into a My Block. Be sure to clearly label each letter so you can tell them apart.

writing a message

To write your automated message, assemble a string of My Block letters. For example, the program in Figure 4-9 signals the word *LEGO*. The program includes two blocks that flash a yellow light at the beginning of the program to indicate that the message is starting over. This indicator is optional but helps recipients decode the message.

This final program, which uses layers of My Blocks to make the program quick and efficient, includes the following:

1. **Loop block**   This repeats the Morse code message infinitely.

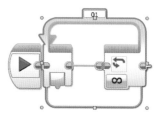

Attach a Loop block after the Start block.

2. **Brick Status Light block**   This sets a yellow flashing light to indicate the beginning of a message. Every time the loop begins the message, the yellow light lets the recipient know that the message is repeating.

Place the Brick Status Light block inside the loop. Set **Brick Status Light** to **On**, Color to **1**, and **Pulse** to **True**.

3. **Wait block**   This flashes the yellow light for 3 seconds before the message starts.

Place the Wait block after the Brick Status Light block. Set **Wait** to **Time** and **Seconds** to **3**.

4-7. **Letter My Blocks**   These spell out letters in Morse code.

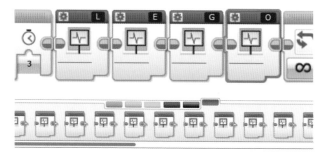

The My Blocks you created earlier for all the letters in the alphabet appear under the My Blocks tab at the bottom of the screen. The letter each block produces appears in the block's upper-right corner. Use the letter blocks to spell out the message you want to display.

FIGURE 4-9: A program for repeatedly flashing Morse code for LEGO

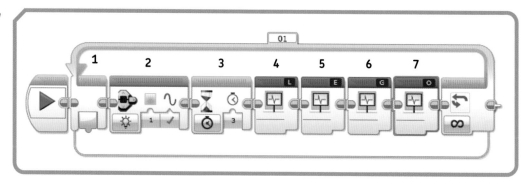

using the automated message transmitter

Your Automated Message Transmitter is ready to put into action. Try using it as a holiday decoration that hangs in a window of your home. Display greetings such as *TRICKOR-TREAT* for Halloween, *LETSEATTURKEY* for Thanksgiving, and *HAPPYNEWYEAR* to celebrate New Year's Day. Wait and see which of your neighbors figures it out. Or you can add a LEGO touch to your holiday table setting, as in Figure 4-10, where I've added the Morse code greeting *HAPPYEARTHDAY* to a table centerpiece composed of my mineral collection.

Which of your guests can figure out the program?

what you learned

In this chapter, you learned how to use Morse code to send messages, an especially useful skill in certain emergencies. You built both manual and automated signaling devices from the MINDSTORMS EV3 Intelligent Brick. You worked with My Blocks in the MINDSTORMS programming environment, an advanced technique for creating efficient software programs. In Chapter 5, you'll continue to use Morse code to send messages, this time with infrared light to mask your communications.

FIGURE 4-10: *An automated message adds interest to tabletop decorations, such as this collection of minerals displayed for Earth Day.*

5

infrared message transmitters

Infrared light exists outside the range of human vision. It can be especially useful if you need to communicate stealthily or if you're moving around at night and want to avoid detection. Or maybe you're interested in countering the actions of other stealthy people. Either way, using it offers advantages. In this chapter, you'll create two LEGO-based devices that work with infrared light: an infrared version of the automated Morse code transmitter you built in Chapter 4, to keep your messages secret, and an infrared receiver, which converts invisible infrared signals from the MINDSTORMS EV3 Infrared Beacon into visible ones.

working with infrared light

Infrared light is a range of colors that exist beyond the color red in the electromagnetic spectrum, shown in Figure 5-1.

Your eye can't see those colors, but certain electronic devices can make them visible. For example, the silicon-based photodetectors found in common digital cameras can capture infrared wavelengths. However, camera manufacturers usually put a glass filter inside the camera to block infrared light and avoid making images that look different from what people see. For the project in this chapter, you'll need an imaging device with no filter or a filter that is removable. Here are a few options:

Some smartphones Some phones have cameras with infrared filters, and some don't. Some phones even have two cameras—a front-facing camera and a back-facing camera—one with an infrared sensor and the other without. For example, an iPhone SE will capture infrared light on the front-facing camera but not on the back-facing camera. Test your phone camera by shining the EV3 Infrared Beacon at the camera. Infrared light likely shows up as purple on a phone without an infrared filter.

Night vision equipment These days, you can get night vision monoculars (Figure 5-2) for as little as $100. The infrared light will be much brighter when viewed through a device like this than through a smartphone.

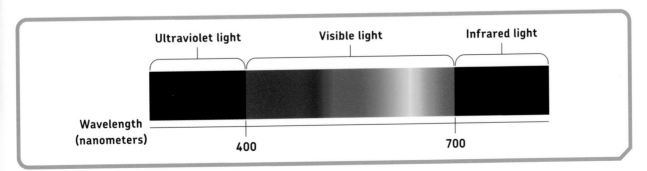

Ultraviolet light   Visible light   Infrared light

Wavelength (nanometers) — 400 — 700

FIGURE 5-1: Infrared light has a wavelength longer than wavelengths in the spectrum of visible light.

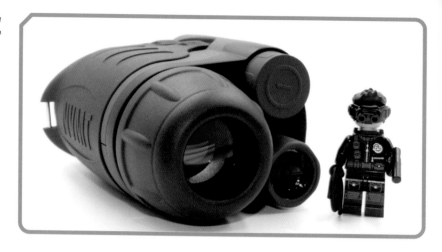

Camcorders   Some camcorders have an IR mode or night mode setting, which, when activated, either engages a mechanism inside the camcorder to pull the infrared-blocking filter out of the way or uses software processing to let the infrared appear in the recorded images.

Some single-lens reflex cameras   Some of these cameras have removable infrared-blocking filters in front of the reflex mirrors.

what you'll need

To build the projects in this chapter, you'll need the following:

* Any infrared viewer, like the ones described in the preceding list.

* A Grove LED Socket Kit, used to connect an external LED to the EV3 Intelligent Brick. You can find this part at Seeed Studio (*https://www.seeedstudio.com/*) or from electronic parts distributors such as RobotShop (*https://www.robotshop.com/*). These kits are sold with a particular color LED, with a name like *Grove Green LED*. You can get any color, since you'll be replacing the visible light LED with an infrared LED.

* A holder for the Grove LED Socket Kit, such as a 3 × 3 Grove Wrapper (also made by Seeed Studio). Openings in the back of the Grove Wrapper let you mount it to LEGO pieces via Technic pins 1/2.

* A Grove Sensor Adapter that allows the EV3 Intelligent Brick to control Grove devices. This is available from Mindsensors.com (*http://www.mindsensors.com/*).

* An infrared LED at a wavelength of 940 nm. This is available from the online electronics store Jameco Electronics (*https://www.jameco.com/*) as part number 2275839. Any infrared LED with a 3 mm or 5 mm base and a wavelength between 700 nm and 1000 nm will work, though you should make sure it has a light intensity of at least 50 milliwatts per steradian (mW/st). Light intensity is sometimes alternately listed in terms of millicandela (mcd); you'll want at least 100 mcd with a viewing angle of no more than 30 degrees.

You'll also need the LEGO components in Figure 5-3, which include the same parts as those you used in Chapter 1 to mount the EV3 Infrared Sensor to the side of the EV3 Intelligent Brick.

Although certain parts aren't found in the MIND-STORMS EV3 #31313 set, you could replace the locking turntable with a Technic pin and liftarm design using parts from that set. This alternate approach is described in Chapter 1.

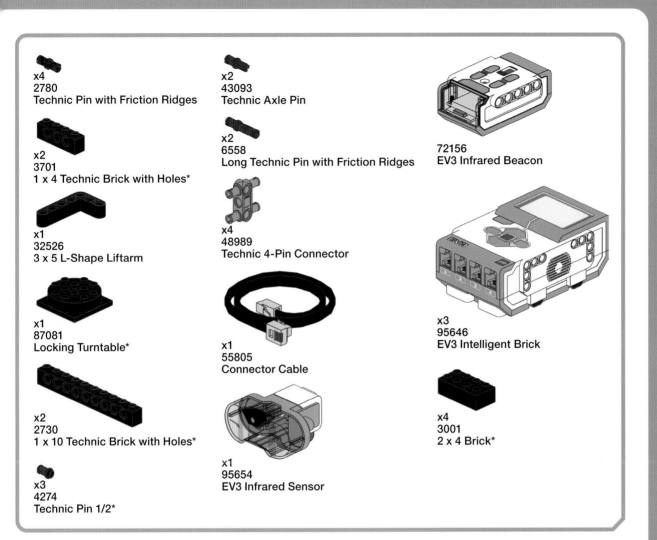

x4
2780
Technic Pin with Friction Ridges

x2
3701
1 x 4 Technic Brick with Holes*

x1
32526
3 x 5 L-Shape Liftarm

x1
87081
Locking Turntable*

x2
2730
1 x 10 Technic Brick with Holes*

x3
4274
Technic Pin 1/2*

x2
43093
Technic Axle Pin

x2
6558
Long Technic Pin with Friction Ridges

x4
48989
Technic 4-Pin Connector

x1
55805
Connector Cable

x1
95654
EV3 Infrared Sensor

72156
EV3 Infrared Beacon

x3
95646
EV3 Intelligent Brick

x4
3001
2 x 4 Brick*

FIGURE 5-3: LEGO parts used for experiments with infrared light. Parts marked with an asterisk are not in the MINDSTORMS EV3 #31313 set.

the covert Morse code transmitter

An automated Morse code transmitter, like the one built in Chapter 4, that uses invisible infrared light would be a great innovation on the espionage concept of the dead drop. A *dead drop* is a designated place for spies to leave messages to each other. Spies can use it to communicate without having

to meet in person. But using the dead drop can be risky if the spies have to physically touch the dead drop location point.

A safer way to leave a message would be to program the message in Morse code with a blinking infrared LED and set the light in a window. The recipient of the message can view the window from a safe distance, such as from the inside of a car. To see the infrared, they could use a night vision monocular or record the message with a smartphone or camcorder.

building the covert Morse code transmitter

To build the Covert Morse Code Transmitter, you'll connect an infrared LED to the EV3 Intelligent Brick with the Grove Sensor Adapter and Grove LED Socket Kit, shown in Figure 5-4.

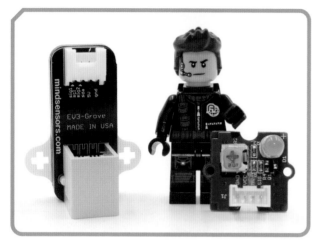

FIGURE 5-4: *The Grove Sensor Adapter (at left) allows the EV3 Intelligent Brick to interface with various Grove devices, such as the LED Socket Kit (at right).*

First, replace the visible color LED of the Grove device in Figure 5-4 with an infrared LED, such as the one shown in Figure 5-5.

FIGURE 5-5: *Replace the visible-wavelength LED with an infrared LED, such as this device from Jameco (part number 2275839).*

Note that the new infrared LED you purchased has one lead that is longer than the other. The longer lead is the positive connection, which goes into the positive socket, labeled + on the Grove LED Socket Kit. Another way to distinguish polarity is that some LEDs indicate negative polarity with a straight edge in the LED's rounded base—the lead by the straight edge is the negative connection.

Before installing your new infrared LED, trim its leads to about the same length as those from the visible LED you removed from the Grove LED Socket Kit, but keep track of which lead is positive. Then insert it into the Grove socket.

Next, attach the modified Grove LED Socket Kit and Grove Sensor Adapter onto the EV3 Intelligent Brick, as shown in Figure 5-6.

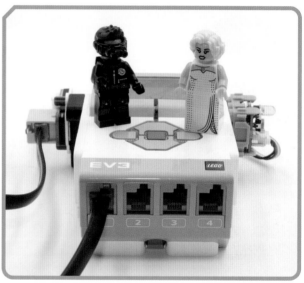

FIGURE 5-6: *The Grove Sensor Adapter is on the left side of the EV3 Intelligent Brick. The Grove LED Socket Kit, held in a Grove Wrapper, is on the right.*

Follow the building instructions to attach the Grove devices.

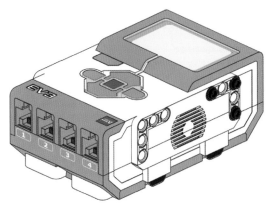

1

2

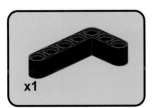

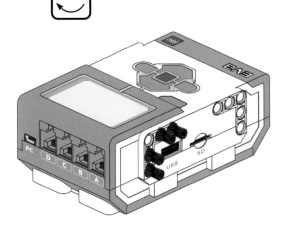

3

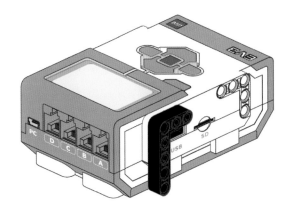

x2

4

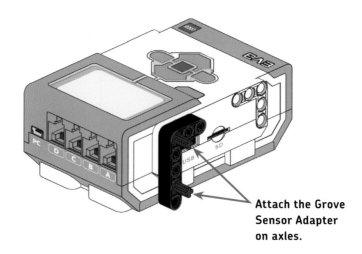

Attach the Grove
Sensor Adapter
on axles.

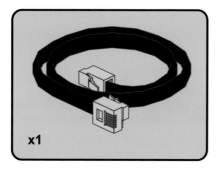

x1

5

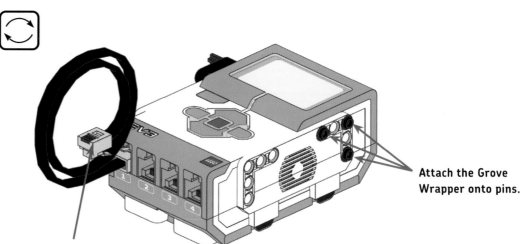

Attach the Grove
Wrapper onto pins.

Connect the end of
the cable to the Grove
Sensor Adapter.

writing the software for the covert Morse code transmitter

The software for sending infrared Morse code messages resembles the program for the Automated Message Transmitter in Chapter 4, except that you'll control the infrared LED rather than lights and sound on the EV3 Intelligent Brick.

To control the Grove LED, download the Grove Sensor Adapter programming block available from Mindsensors.com at *http://www.mindsensors.com/ev3-and-nxt/20-grove-sensor-adapter-for-ev3-or-nxt*. This will download a file onto your computer called *GroveSensorAdapter.ev3b*. Import it into the MINDSTORMS EV3 programming environment by going to the toolbar and selecting **Tools ▸ Block Import**. Select the *GroveSensorAdapter.ev3b* Grove block from the location where you downloaded it. The Grove Sensor Adapter block will now appear in the Sensor tab, and you can drag it into programs like any other MINDSTORMS EV3 programming block.

As you did in Chapter 4, you'll build separate programs for displaying the Morse code dashes and dots, and then turn those programs into My Blocks that you can use to create programs representing each letter in the alphabet. You can then make My Blocks of these programs and write your own message with them.

building the dot and dash programs

Figure 5-7 shows the dot and dash programs.

The dot program will create a short flash of the infrared LED, and the dash program will create a long flash. To build the programs, use the following blocks:

1. **GroveSensorAdapter block**   This turns on the Grove LED.

Go to the Sensors tab at the bottom of the screen and click the GroveSensorAdapter block. Place the Grove block next to the program Start. Set **Mindsensors GroveSensorAdapter** to **Digital**, **Port** to **1**, **I2CAddr** to **66**, and **Digital Output On?** to **True**.

The Grove Sensor Adapter's digital mode lets it turn on or off the device connected to a port—in this case, the Grove LED Socket Kit connected to port 1. The drop-down menu command for Digital Output On? tells the LED to turn on (for the true setting) or turn off (for the false setting). You'll use this on-and-off combination to blink the infrared LED.

FIGURE 5-7: The dot and dash programs

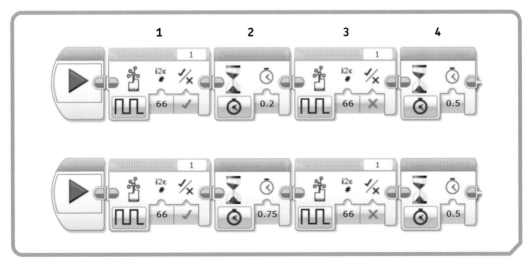

2. **Wait block**   This keeps the infrared LED on for 0.2 seconds for a dot or 0.75 seconds for a dash.

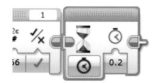

   Select a Wait block from the Flow Control tab and place it after the GroveSensorAdapter block. Set **Wait** to **Time** and **Seconds** to **0.2** for a dot or **0.75** for a dash.

3. **GroveSensorAdapter block**   This turns off the infrared LED.

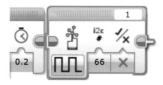

   Place a GroveSensorAdapter block after the Wait block. Set **Mindsensors GroveSensorAdapter** to **Digital**, **Port** to **1**, **I2CAddr** to **66**, and **Digital Output On?** to **False**.

4. **Wait block**   This pauses for 0.5 seconds before the next dot or dash.

   For the Wait block, set **Wait** to **Time** and **Seconds** to **0.5**.

FIGURE 5-8: Programs for flashing the letters A and B in Morse code

building the alphabet blocks

To create blocks that flash each letter of the alphabet in Morse code, you must first create My Blocks for the dot and dash subroutines. Follow the steps in Chapter 4 if you're not sure how to do this. Call your new My Blocks **IR_dot** and **IR_dash**.

Next, use these new blocks to write programs for each Morse code letter. (Consult Chapter 4 if you need a Morse code refresher.) Figure 5-8 shows the programs for the letters A and B.

Because you won't be able to insert a different-colored flash to mark the end of each letter, you'll separate the letters with a time delay instead.

Use the following process to create each letter. These specific blocks create the letter A:

1. **My Block IR_dot**   This blinks a dot by using the dot My Block.

   Drag the IR_dot block from the My Blocks tab at the bottom of the screen and place it next to the program Start.

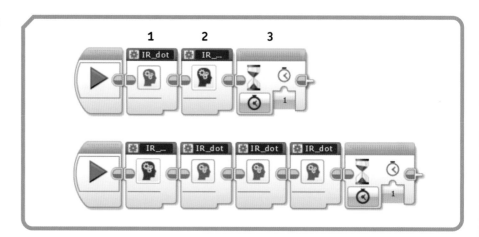

2. **My Block IR_dash**   This blinks a dash by using the dash My Block.

Place the IR_dash My Block after the dot.

3. **Wait block**   This pauses for 1 second before flashing the next letter. This delay gives the message viewer a pause to get ready for the next letter.

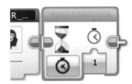

Set **Wait** to **Time** and **Seconds** to **1**.

Make My Blocks for all the letters of the alphabet, following the procedure for creating My Blocks described in Chapter 4.

writing a message program

Once you have your Morse code alphabet, compose a message by stringing together letter My Blocks. For example, Figure 5-9 creates a program for flashing the word *LEGO*.

Use the following blocks to create your message program:

1. **Loop block**   This repeats the Morse code message infinitely.

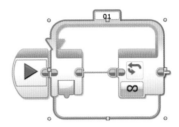

Insert the Loop block after the program Start.

2. **Brick Status Light block**   This turns off the EV3 Intelligent Brick's front panel lights that would ordinarily flash during operation. This flashing light would ruin the secret transmission.

Select the Brick Status Light block from the Action tab, place it inside the loop, and set **Brick Status Light** to **Off**.

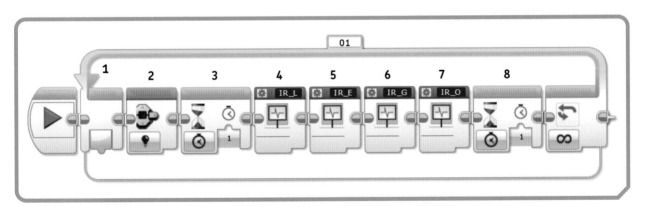

FIGURE 5-9: A program that flashes the word LEGO *in infrared light Morse code*

3. **Wait block**   This pauses for 1 second before starting the message so that the recipient knows the message is starting over.

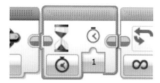

Place the Wait block after the Brick Status Light. Set **Wait** to **Time** and **Seconds** to **1**.

4-7. **My Blocks**   Blink the Morse code letters for your message.

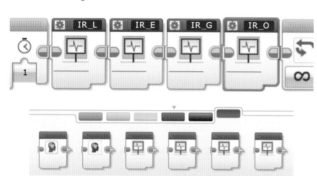

Go to the My Blocks tab at the bottom of the screen to select the letters for your message. Line up the letters after the Wait block.

8. **Wait block**   This pauses for 1 second before starting the message over again.

Place the Wait block after your letter My Blocks. Set **Wait** to **Time** and **Seconds** to **1**.

viewing your covert Morse code

The infrared LED will start blinking as soon as you've downloaded your Morse code message onto the EV3 Intelligent Brick—but, of course, you won't see the blinking LED since it's infrared. To see your message, you'll have to use one of the devices discussed earlier in this chapter. Viewing the infrared LED with these devices becomes easier in darker environments, but you can still see the beam in a room with the lights on. Figure 5-10 shows the image viewed through an iPhone SE. You can record a video of the blinking light on a smartphone or camcorder to decode or replay later as needed.

The message is easiest to see on a night vision monocular. Figure 5-11 shows an example using the monocular from Figure 5-2. Viewed through a night vision monocular, the blinking LED shows up quite clearly in white. The night vision monocular also illuminates the area surrounding the EV3 Intelligent Brick.

If you're having a problem seeing the infrared LED blinking, you've most likely inserted the LED into the Grove LED Socket in the wrong positive/negative orientation. You can solve this by pulling out the infrared LED, rotating it, and plugging it back in. If this doesn't work, remove the infrared LED and temporarily replace it with the visible wavelength LED that came with the Grove LED Socket. If the LED still doesn't blink, the problem is likely with the software.

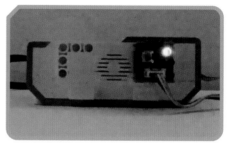

FIGURE 5-10: *The blinking LED shows up as a purple light when viewed through an infrared-capable camera on a smartphone.*

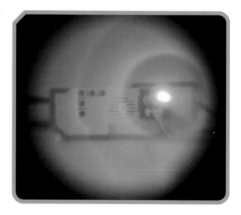

FIGURE 5-11: *Through a night vision device, the infrared LED appears as a white pulse.*

infrared Morse code receiver

Another way you could view an infrared message is to build a receiver that converts infrared blinking into visible light and audible tones. In this section, you'll build such a receiver by making use of the EV3 Infrared Sensor and the EV3 Intelligent Brick. You'll use the EV3 Infrared Beacon as a remote to manually flash Morse code infrared messages. The receiver's sensor should detect the signals and play the message.

You could use this device to receive secret messages transmitted by someone who has the EV3 Infrared Beacon. With the Infrared Morse Code Receiver, you can send and receive messages tapped out in real time.

FIGURE 5–12: *The Infrared Morse Code Receiver*

building the infrared Morse code receiver

The hardware for the Infrared Morse Code Receiver, pictured in Figure 5-12, is the same as for the infrared sensor head in Chapter 1, so you can consult those building instructions to put it together.

The sensor will detect Morse code messages sent from the remote control.

writing the infrared Morse code receiver software

Figure 5-13 shows the software to make a receiver out of the EV3 Intelligent Brick. The Infrared Morse Code Receiver uses the EV3 Infrared Sensor to detect when a button is pressed on the EV3 Infrared Beacon. When the button is pressed, the EV3 Intelligent Brick will sound a tone and light up the front panel lights. You'll use the same light and sound activation inside the Switch block as you did for the Manual Message Transmitter of Chapter 4.

FIGURE 5–13: *The Infrared Morse Code Receiver program is based on an Infrared Sensor Switch block.*

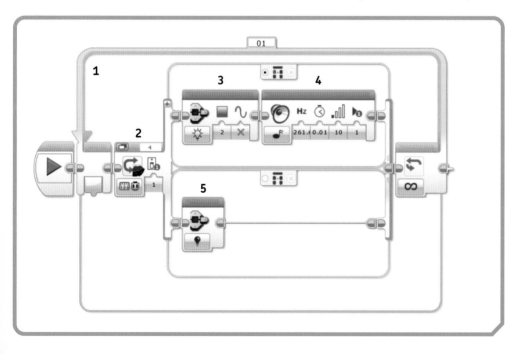

Build the program with the following blocks:

1. **Loop block**   This infinitely waits for a button to be pressed on the EV3 Infrared Beacon.

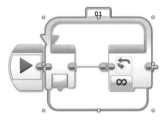

Place a Loop block after the program Start.

2. **Switch block**   This creates true and false paths. If someone presses a button on the EV3 Infrared Beacon, the true path should execute. Otherwise, the false path should run.

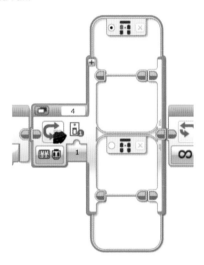

Install the Switch block inside the Loop. Set it to **Infrared Sensor–Measure–Remote**, **Port** to **4**, and **Channel** to **1**. Make sure the EV3 Infrared Beacon is also set for Channel 1 by using the slide switch on the front of the device.

3. **Brick Status Light block**   This activates a red light on the front panel of the EV3 Intelligent Brick. This light will stay on for as long as the EV3 Infrared Beacon's button is pressed.

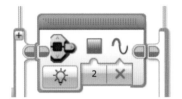

Place a Brick Status Light block inside the true path of the Switch. Set **Brick Status Light** to **On**, **Color** to **2**, and **Pulse** to **False**.

4. **Sound block**   This plays a sound through the EV3 Intelligent Brick's speaker for as long as the EV3 Infrared Beacon's button is pressed.

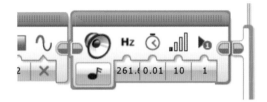

Place a Sound block after the Brick Status Light. Set **Sound** to **Play Tone**, **Frequency** to **261.626**, **Duration** to **0.01**, **Volume** to **10**, and **Play Type** to **1**.

5. **Brick Status Light block**   This turns off the light on the EV3 Intelligent Brick's front panel. This block begins the false path of the Switch block, which runs when no button is pressed on the EV3 Infrared Beacon.

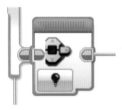

Place a Brick Status Light block in the false path of the Switch block and set **Brick Status Light** to **Off**.

using the infrared Morse code receiver

After downloading the program onto the EV3 Intelligent Brick, you're ready to send messages across a room. Place the receiver with the EV3 Infrared Sensor rotated to face roughly in the direction from which you intend to transmit. Press any of the four buttons on the EV3 Infrared Beacon to activate the LED front panel lights and speaker of the EV3 Intelligent Brick. Experiment with how far away you can be and still successfully transmit messages, or test whether you can send messages through window glass.

what you learned

In this chapter, you worked with infrared light, which is invisible to the human eye. You used infrared light sources to covertly send messages. Only someone who knows how to detect light from the infrared spectrum will be able to decode your signals. You also learned how to add a wide array of devices to MINDSTORMS EV3 elements with the Grove system. Later chapters will explore more capabilities of Grove devices.

6

the automatic insect trapper

Entomology (the scientific study of insects) is a vital field of research. An amateur scientist like yourself can contribute to this field by studying the insects in your area. In this chapter, you'll build an automated trap that will capture nocturnal insects but not kill them.

The Automatic Insect Trapper attracts insects to a box by using ultraviolet (UV) light, an established technique used by professional entomologists. Usually, though, entomologists funnel the insects into a container that kills them. Instead, you'll design the Automatic Insect Trapper to automatically close the lid on the box after the insects have been baited with the UV light. You can then let the insects go by simply opening the box's lid.

SAFETY WARNINGS

Working with electricity and UV light poses some safety hazards. You'll make an electrical connection to *household alternating current* (that is, a 120 VAC or, in certain countries, a 240 VAC wall outlet), which can pose hazards for electrocution or fire. The connections and switch described in this chapter are no more dangerous than household appliances, but take the following precautions:

* Don't modify or cut into 120 VAC (or 240 VAC) cables, wires, or connectors. This can expose the wires to contact with lethal voltage.

* If you install this project outside, make sure the electrical parts remain dry. Don't leave electrical cables or parts outside in the rain or other moisture.

* Plug the experiment into a ground-fault interrupted circuit rated for outdoor use.

* Keep electrical connections far away from any swimming pool, spa, or water garden.

* Don't leave the experiment unattended for a long period.

If misused, UV light can damage your eyes or skin, with greater risks at certain wavelength ranges. There are three classes of UV light sources, with increasing levels of hazard. Even though you'll be using UV-A, the safest class, you should take the following precautions:

* Don't let kids play with a UV light source.

* Don't stare into a UV light source.

* Don't hold a UV light source close to your skin.

In addition, some UV light sources can get hot and pose a fire hazard if left on inside a closed space, which is why this project uses an LED UV-A light bulb.

building the automatic insect trapper

The Automatic Insect Trapper, shown in Figure 6-1, attracts insects with a UV lamp placed inside a hinged box. You'll keep the lid open by using a string connected to an *electromagnet*, which is a magnet that you can turn on or off with an electrical current. After two hours, the EV3 Intelligent Brick will turn off the electromagnet, causing the lamp to turn off and the lid to drop.

You'll build three subsystems for this project: the box, the UV lamp, and the lid release.

what you'll need

Figure 6-2 shows the LEGO parts used to build the insect trap.

In addition to these LEGO parts, you'll need the following:

* Large box with a hinged lid, such as the Plano Large Hinged Storage Box (*https://www.planomolding.com /storage/*)

* LED UV-A light bulb, such as the Sunlite A-19 bulb available at Amazon (*https://www.amazon.com/*)

* Lamp, such as a small desk lamp, for the UV-A light bulb

* Grove Electromagnet and Grove Wrapper, available from Seeed Studio (*https://www.seeedstudio.com/*), RobotShop (*https://www.robotshop.com/*), or Amazon (*https://www .amazon.com/*). A four-wire cable is part of the purchase when you buy the Grove Electromagnet.

* Grove Sensor Adapter from Mindsensors (*https://www.mindsensors.com/*)

* Breadboard Connector Kit from Mindsensors to use as a breakout board. This kit includes three parts: a circuit board, an EV3 connector, and a header. You'll be working with the circuit board and EV3 connector but not the header.

* A 0.75-meter or longer Custom Cut Flexi Cable from Mindsensors

* A 2-meter or longer Custom Cut Flexi Cable from Mindsensors

* Two jumper wires of different colors, preferably one red and one black

FIGURE 6-1: *The Automatic Insect Trapper catches insects attracted to an ultraviolet light.*

x6
2780
Technic Pin with Friction Ridges

x4
4274
Technic Pin 1/2*

x2
43093
Technic Axle Pin

x2
32525
1 x 11 Technic Liftarm

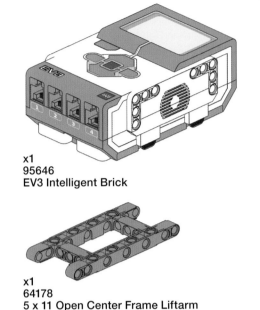

x1
95646
EV3 Intelligent Brick

x1
64178
5 x 11 Open Center Frame Liftarm

FIGURE 6-2: *All the LEGO parts used to build the Automatic Insect Trapper are in the MINDSTORMS EV3 #31313 set, except for the Technic pins 1/2, as marked by an asterisk.*

* A 120 VAC relay, such as the IoT Control Relay made by Digital Loggers (*https://www.digital-loggers.com/*). If your country uses 240 VAC power, you can use a power converter to adapt to this 120 VAC device.

* Cyanoacrylate glue, such as Super Glue, Gorilla Glue, or Krazy Glue

* String

* A washer and a steel tool (such as a universal joint socket for a ratchet tool) that will stick to a magnet

* Zip ties

* A 120 VAC outdoor extension cord

* Tools for making electronic connections, including a soldering iron, solder, diagonal cutting pliers, and wire strippers

* For optional windows for the box, plexiglass sheets of 2 mm (0.08 inch) thickness. You can cut up a 20 cm × 25 cm sheet, available at most hardware stores, into 8 cm × 12 cm sections, or you can buy precut sections (meant to cover photographs) at craft stores or online at Amazon.

* For optional modifications to the box to install windows, a hand jab saw or electric rotary tool

modifying the box

Start by setting up the box you'll use to trap bugs. It should be large, with a hinged lid that is heavy enough to slam shut if dropped, but not so heavy that opening the lid takes much force. Figure 6-1 shows a Plano Large Hinged Storage Box.

Make the following modifications to the box:

1. Pull a string through the hole in the front of the lid that's typically used for a lock. To keep the string from falling out, tie a washer to one end of it. Tie a steel tool to the other end of it. This tool can be anything made of steel, allowing it to stick to a magnet. For example, Figure 6-1 uses a universal joint socket to stick to the magnet. Universal joint sockets, used with ratchet wrenches, are available at hardware stores.

2. If you want to be able to see your bugs in the box without opening it, cut rectangular openings into the box lid with a hand jab saw or an electric rotary tool and cover them with plexiglass to create windows. You can use cyanoacrylate glue to hold the plexiglass in place.

3. With a jab saw or electric rotary tool, cut a 1 cm × 1 cm square into the rim of the box so you can pass two cables into the interior of the box. One of these cables will be a 120 VAC extension cord; the other will be an EV3 cable (the longer Mindsensors Custom

Cut Flexi Cable) that makes a connection to the electromagnet lid-release mechanism. This square notch, with cables passing through, can be seen in Figure 6-1 on the left rim of the box. If you don't cut this notch, your trap will still work, but you may lose a few of the smaller insect specimens that can crawl out if the cables prevent the lid from fully closing.

setting up the UV lamp

Next, let's set up the UV lamp. Insects are the most sensitive to light that has a wavelength of about 365 nm, which lies within the UV-A range of the ultraviolet spectrum. The Sunlite bulb (Figure 6-3) provides light at that wavelength.

FIGURE 6-3: The Sunlite UV-A bulb has a base that can be screwed into any ordinary lamp.

Attach this bulb to any ordinary lamp that fits inside the box when the lid is shut, such as the flexible-neck desk lamp used in Figure 6-1. Place the lamp so the ultraviolet bulb faces upward, enabling insects to easily see the light. You'll later plug the lamp's power connection to the relay device described in the next section.

powering the light

You'll turn the lamp on and off with an electrical control provided by a relay. A *relay* is a device that uses an electronic signal to activate a different electrical circuit. You'll use the IoT Control Relay, pictured in Figure 6-4, so that the EV3 Intelligent Brick can control whether the ultraviolet light is on or off.

You'll need to build a cable connection to the relay that can adapt to the EV3 Intelligent Brick. This custom cable connection, also shown in Figure 6-4, consists of a *breakout board*, a 0.75-meter (or longer) Mindsensors Custom Cut Flexi Cable, and red and black wire leads. A breakout board, such as the Mindsensors Breadboard Connector Kit, allows access to the internal pins on a connector, such as the connectors built into the EV3 Intelligent Brick. The Mindsensors Custom Cut Flexi Cable serves the same function as the LEGO 55805 Connector Cables that come in the MINDSTORMS EV3 #31313 set, but it's longer and more flexible. You'll attach one of these custom cables to a breakout board in order to connect the IoT Control Relay to the EV3 Intelligent Brick.

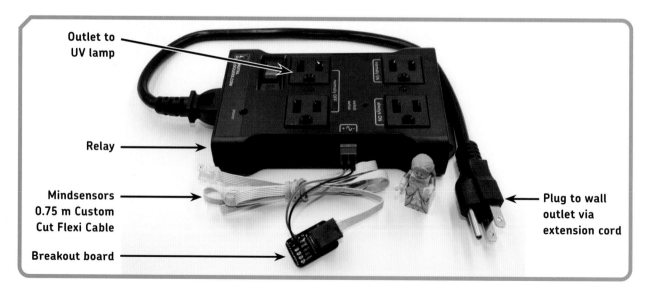

FIGURE 6-4: The relay switches the UV lamp on and off by using a control signal applied to the red and black wire pair.

The breakout board connects the six wires inside the Mindsensors Custom Cut Flexi Cable to solder pads. Each solder pad is labeled on the breakout board and matches the corresponding wire inside the Mindsensors cable (these wires would be visible to you only if you cut open the cable). You'll be using the breakout board to connect the cable to the positive and negative connections on the IoT Control Relay.

FIGURE 6-5: Solder wires onto AN IN and GRND on the breakout board. In an upcoming step, you'll connect the other end of the wires to the screw terminals on the IoT Control Relay.

soldering the jumper wires to the breakout board

Solder a pair of jumper wires onto the breakout board, one at the pin labeled AN IN and one at the pin labeled GRND, as shown in Figure 6-5. There are two GRND pins on the breakout board—use the one next to the AN IN pin.

Use different colors for the jumper wires so you can tell at a glance which wire is the positive connection and which is the negative connection. For example, in Figure 6-5, a red wire is soldered to AN IN for the positive connection, and a black wire is soldered to GRND for the negative connection.

soldering the EV3 connector to the breakout board

Next, install and solder the EV3 connector onto the break-out board—this connector came with the Mindsensors Breadboard Connector Kit. The EV3 connector fits into the breakout board only one way, such that the solder leads and mounting posts all line up with holes in the breakout board. The connector's opening, where the cable plugs in, faces away from the labels printed on the breakout board.

FIGURE 6-6: You connect the control signal to the IoT Control Relay via a pull-out green connector.

connecting the jumper wires to the relay

Connect the other end of the jumper wires into the IoT Control Relay, as shown in Figure 6-5. To make this connection, pull the green connector block out of the side of the relay. You'll then see two screws to loosen on the top of the green connector so that the wire jumpers can be installed. Place the wire jumpers into the side holes of the green connector and then tighten the two screws on top of the green connector, as shown in Figure 6-6. Press the green connector back into the IoT Control Relay.

Set the IoT Control Relay aside for now. You'll make the power and cable connections after you've taken the Automatic Insect Trapper outdoors.

building the lid release

The lid release allows you to keep the box open by tugging on a string until the time comes to release it. When the string is released, the box's lid will slam shut. To grip the string, you'll use a Grove Electromagnet, shown in Figure 6-7, along with the Grove Sensor Adapter.

A digital control signal provided by an EV3 program will deactivate the electromagnet when it's time to close the Automatic Insect Trapper, releasing the string and slamming the box shut.

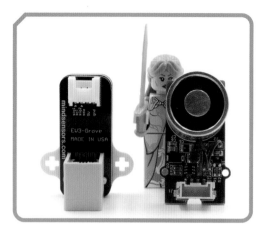

FIGURE 6-7: The Grove Electromagnet, at right, can be turned on and off with the Grove Sensor Adapter, at left.

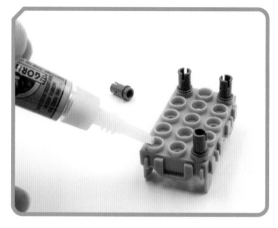

FIGURE 6-8: Prepare a Grove Wrapper to hold the Grove Electromagnet by gluing Technic pins 1/2 into the four corners on the back of the Grove Wrapper.

gluing technic pins to the Grove Wrapper

To strongly fasten the Grove Electromagnet to a LEGO lift-arm, glue four Technic pins 1/2 into the four corner attachment points of the Grove Wrapper, as shown in Figure 6-8.

After the glue has dried, press the Grove Electromagnet into the front of the Grove Wrapper, as shown in Figure 6-9. This attachment is strong enough to hold the Grove Electromagnet in place—there's no need for glue on the front mounting.

connecting the Grove Electromagnet and Grove Sensor Adapter

Connect the Grove Electromagnet to the Grove Sensor Adapter by using a structure made from a 5 × 11 open center frame liftarm, as shown in Figure 6-9.

Follow these building instructions to make the structure.

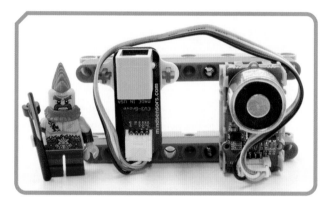

FIGURE 6-9: Attach the Grove Electromagnet and Grove Sensor Adapter onto liftarms.

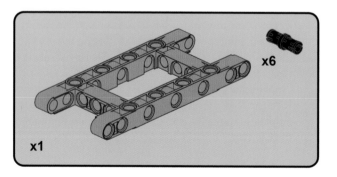
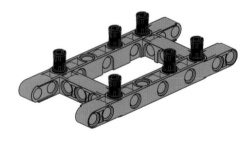

1

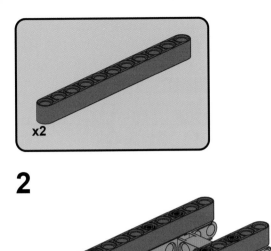

2

3

Attach the Grove
Sensor Adapter
on axles.

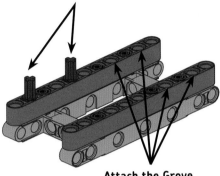

Attach the Grove
Wrapper at four points.

After you've built the LEGO structure and installed the Grove Sensor Adapter and Grove Wrapper, attach the wire cable between the Grove Sensor Adapter and Grove Electromagnet, as shown in Figure 6-9. This wire cable should have been included in the same package as the Grove Electromagnet.

writing the software

Before putting all the parts together, write the program that will control the lamp and the electromagnet. Figure 6-10 shows the program to operate the Automatic Insect Trapper. This program should keep the trap's lid open, turn on the UV lamp for a certain amount of time, and then close the lid and turn off the UV lamp. Turning off the lamp keeps it from heating up the interior of the closed box.

FIGURE 6-10: The program for the Automatic Insect Trapper

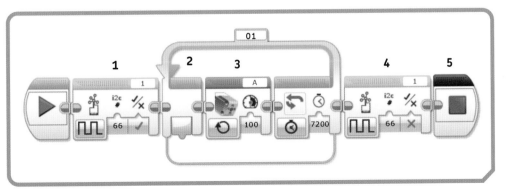

The following blocks will run the Automatic Insect Trapper by controlling the Grove Sensor Adapter and the IoT Control Relay:

1. **GroveSensorAdapter block** This block turns on the Grove Electromagnet by setting the Grove Sensor Adapter to true. You can find instructions to install this block in Chapter 5.

   Select the GroveSensorAdapter block from the Sensors tab and set **Mindsensors GroveSensor-Adapter** to **Digital**, **Port** to **1**, **I2CAddr** to **66**, and **Digital Output On?** to **True**. This configuration of the Grove Sensor Adapter resembles the one in Chapter 5 except that, here, the digital control turns on and off an electromagnet instead of an LED.

2. **Loop block** This keeps the box's lid open for the amount of time specified in the loop. This is a timed loop rather than the infinite loop used in other chapters. You may want to adjust the length of time the box is open. Two hours (or 7,200 seconds) is a good starting point, as shown here:

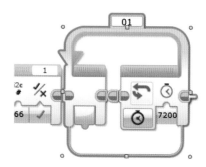

   Select a Loop block and attach it to the GroveSensorAdapter block. Set **Loop** to **Time** and **Seconds** to **7200**.

3. **Medium Motor block** This applies voltage to port A of the EV3 Intelligent Brick in order to activate the IoT Control Relay.

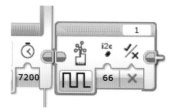

   Place a Medium Motor block inside the Loop and set **Medium Motor** to **On**, **Port** to **A**, and **Power** to **100**.

4. **GroveSensorAdapter block** This drops the lid on the box by changing the GroveSensorAdapter's setting to false, deactivating the Grove Electromagnet.

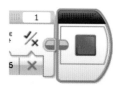

   Add a GroveSensorAdapter block after the Loop block and set **Mindsensors GroveSensorAdapter** to **Digital**, **Port** to **1**, **I2C Addr** to **66**, and **Digital Output On?** to **False**.

5. **Stop Program block** This ends the program. Terminating the program in this way makes sure that the Grove Electromagnet and IoT Relay are turned off.

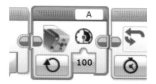

   From the Advanced tab, select a Stop Program block and place it at the end of the program.

connecting the automatic insect trapper

Attach your lid-release mechanism to a tree, fence post, or light post with zip ties, as shown in Figure 6-11. Place the mechanism about 50 cm above the ground or surface where the box will sit.

Place the box below the lid-release mechanism and place the EV3 Intelligent Brick and IoT Control Relay on the floor of the box. Place the UV lamp inside the box so that the bulb can be seen from above the box. Also bring outside the 2-meter Custom Cut Flex Cable and 120 VAC outdoor extension cord that you haven't used yet, along with the items you've built.

Connect the various power and EV3 cables, as diagrammed in Figure 6-12.

These connections consist of the following:

* The 2-meter Mindsensors Custom Cut Flexi Cable between the Grove Sensor Adapter and port 1 of the EV3 Intelligent Brick

* The 0.75-meter Mindsensors Custom Cut Flexi Cable between the breakout board and port A of the EV3 Intelligent Brick

FIGURE 6-11: The lid-release mechanism is attached to a post with zip ties.

* The UV lamp's power cord into a normally OFF outlet on the IoT Control Relay. (Be sure to use one of the normally OFF outlets, not a normally ON or always ON outlet. The normally OFF outlet comes on only when activated by the EV3 Intelligent Brick. It's important for safety to make sure the lamp is not lit when the trap's lid is closed.)

* Power cord of the IoT Control Relay to a 120 VAC extension cord

* Other end of the extension cord to a 120 VAC outdoor outlet

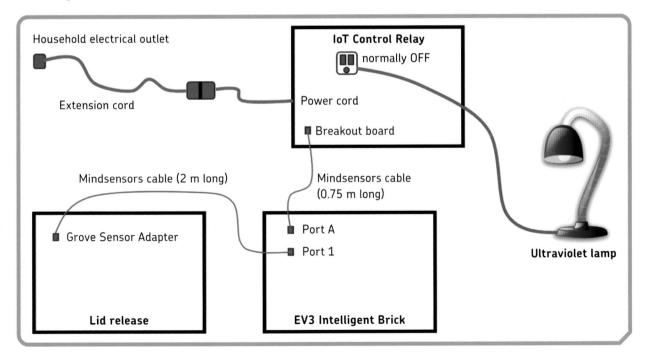

FIGURE 6-12: Connections to parts of the Automatic Insect Trapper consist of power (thick red lines) and Mindsensors control signals (thin green lines).

using the automatic insect trapper

Turn on the EV3 Intelligent Brick and activate the Automatic Insect Trapper program. Both the UV lamp and Grove Electromagnet should now be on. Attach your steel tool–tipped string to the Grove Electromagnet so that the box's lid stays open, as shown in Figure 6-13.

Leave the setup alone and wait for insects to enter the box. After two hours, the box's lid will fall shut, trapping the insects in the box for you to examine when you're ready. Experiment with using the box at different times over the course of the night to see what you capture. (Be aware that around dusk, the trap might fill with mosquitoes. Opening a box full of mosquitoes is an unpleasant experience.) The more interesting insects (such as the ones shown in Figures 6-14 and 6-15) will likely appear later in the night.

To sample a variety of insects at different times of the night, you could change the program to delay the activation of the IoT Control Relay until a certain hour. For such an experiment, though, you should note that the batteries powering the EV3 Intelligent Brick can be drained within a few hours—alkaline batteries should give you 5.5 hours of operation, but not much more.

FIGURE 6-14: *Most of the insects captured were varieties of beetles, like the two shown here.*

FIGURE 6-15: *I caught many moths, which flew away after I left the box open for a while.*

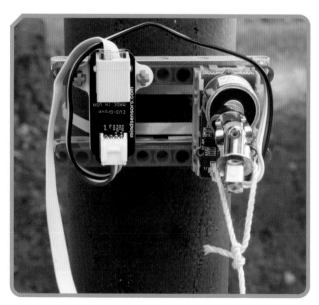

FIGURE 6-13: *After you turn on the Automatic Insect Trapper, the steel tool will stick to the electromagnet to keep tension on the string that holds the lid open.*

what you learned

In this chapter, you built a LEGO-based machine to collect elusive nocturnal insects by combining MINDSTORMS EV3 inventions with a Grove Electromagnet. You also used MINDSTORMS to switch on an electrical device that uses 120 VAC wall outlet power with an IoT Control Relay. You could use this same relay to control other appliances in your home, such as a light source or a coffee pot, for home automation projects.

7

the motion-sensing radar

In this chapter, you'll learn how to add your own sensors to MINDSTORMS EV3 by creating a radar device that can sense motion, as shown in Figure 7-1.

This radar uses the *Doppler effect*, by which objects in motion shift the frequency of a radio wave. If an object such as a human, vehicle, or animal moves through the radar beam, the radar detects its presence.

The Motion-Sensing Radar has two advantages over the standard LEGO-based sensors. First, the radar can see through windows, walls, or doors. So, for example, you can keep the Motion-Sensing Radar indoors with a view through a window to detect whether someone or something is moving outside. Or you could learn whether someone is approaching a closed door. Second, the radar can sense a moving target from farther away than the MINDSTORMS

EV3 Infrared or Ultrasonic Sensors, picking up movement at distances of up to about 10 meters.

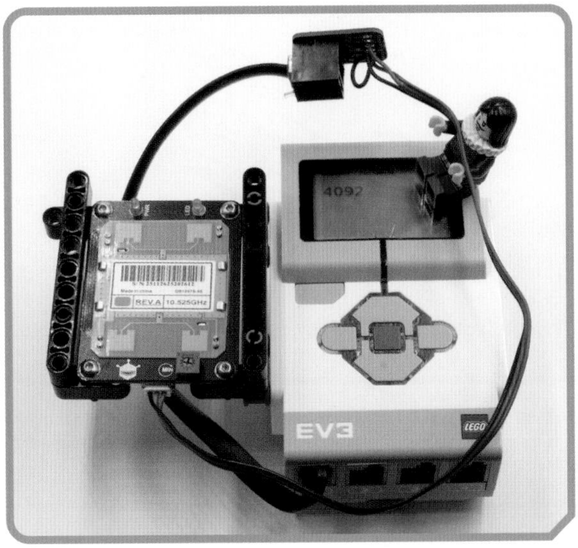

The connection to the EV3 Intelligent Brick you'll create in this chapter can be adapted to any sensor that provides a voltage output as a signal. Such sensors could measure gas concentration, force, distance, proximity, or magnetic fields. You can browse possible sensors by searching for *analog sensor* at sites including DFRobot (*https://www.dfrobot.com/*), RobotShop (*https://www.robotshop.com/*), or Digi-Key (*https://www.digikey.com/*).

building the motion-sensing radar

The Motion-Sensing Radar uses the Digital Microwave Sensor made by DFRobot, shown in Figure 7-2.

FIGURE 7-1: The Motion-Sensing Radar detects moving objects to trigger an alarm on the EV3 Intelligent Brick.

FIGURE 7-2: The Digital Microwave Sensor lights up the yellow LED on the circuit board and creates an electronic pulse when it detects motion.

The yellow part of this device is an antenna that transmits and receives a radio wave of 10.525 gigahertz (GHz). This radio frequency is in the range known as *microwave*, hence the name of the sensor. Microwave ovens, like those commonly found in household kitchens, also use microwaves, but at a lower frequency of 2.45 GHz—a frequency that gets absorbed by the water content in food, making the food heat up. Microwaves at a higher frequency work well for radar. This sensor listens for a reflection of its microwave emission coming from objects within about 10 meters of the device. If the reflection is from a moving target, such as a person walking, then the Digital Microwave Sensor creates an electronic signal. You'll use the EV3 Intelligent Brick to read this electronic signal.

what you'll need

Figure 7-3 shows the LEGO parts used to build the Motion-Sensing Radar. You can find all of these LEGO parts in the MINDSTORMS EV3 #31313 set.

In addition to these LEGO parts, you'll need the following:

* A Digital Microwave Sensor made by DFRobot. This is available from DFRobot (*https://www.dfrobot.com/*), RobotShop (*https://www.robotshop.com/*), Mouser Electronics (*https://www.mouser.com/*), Digi-Key (*https://www.digikey.com/*), or Amazon (*https://www.amazon.com/*). A three-wire cable is part of the purchase when you buy this sensor.

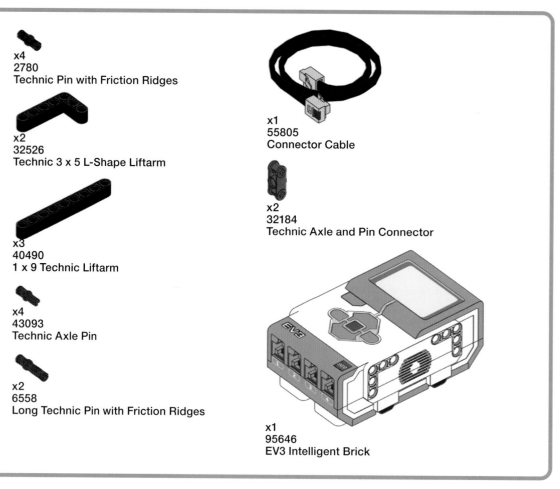

x4
2780
Technic Pin with Friction Ridges

x2
32526
Technic 3 x 5 L-Shape Liftarm

x3
40490
1 x 9 Technic Liftarm

x4
43093
Technic Axle Pin

x2
6558
Long Technic Pin with Friction Ridges

x1
55805
Connector Cable

x2
32184
Technic Axle and Pin Connector

x1
95646
EV3 Intelligent Brick

FIGURE 7-3: *LEGO parts used to build the Motion-Sensing Radar*

A set of four machine screws of M3 size that are 16 mm long or longer, along with matching nuts. (If you have screw sizes that are inch based, you can use 4-40 screws instead.) These screws and nuts are available at hardware stores. Machine screws are available with different head styles. Socket head cap, pan head, or cheese head screws will work.

For the breakout board, a Breadboard Connector Kit from Mindsensors.com (*https://www.mindsensors.com/*)

A jumper wire, preferably black, with a diameter of 0.8 mm (20 American wire gauge) or smaller and a length of about 3 cm

Tools for making electronic connections, including a soldering iron, solder, diagonal cutting pliers, and wire strippers

building the sensor mount

The Digital Microwave Sensor has mounting holes in each of its corners that happen to line up with LEGO spacing. You'll later use M3 screws to attach the sensor to Technic liftarms. First, build the structure for the sensor that will connect to the EV3 Intelligent Brick by performing the following construction steps.

1

x2

2

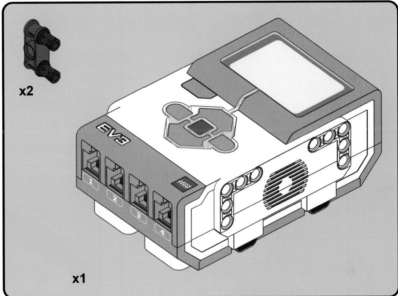

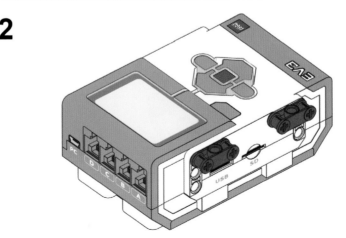

x2

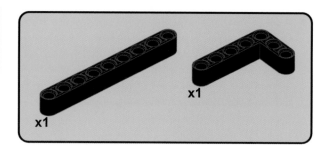

x1   x1

3

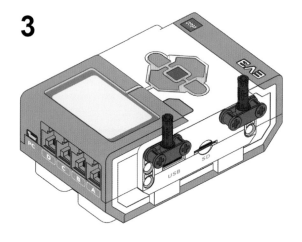

4

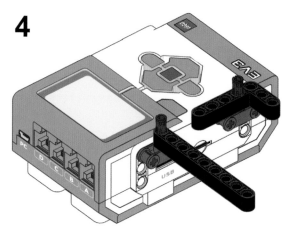

x2

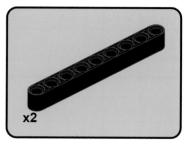

x2

5

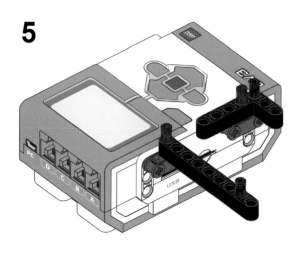

6

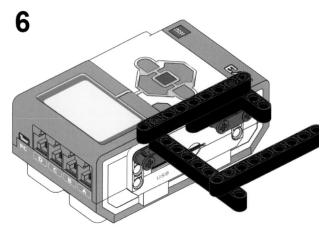

7

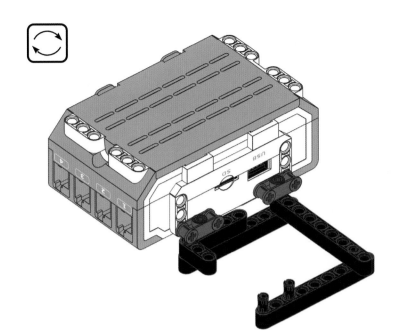

8

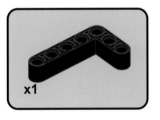

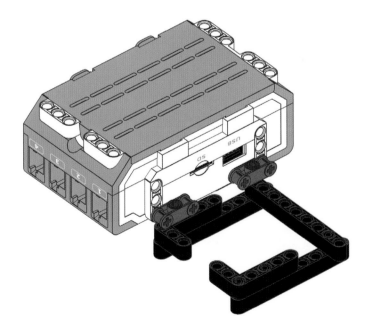

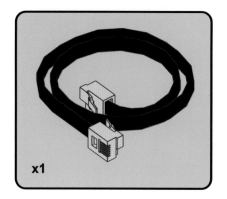

x1

9

Attach the four corners of the sensor with M3 screws and nuts.

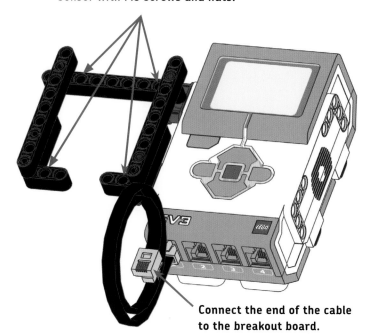

Connect the end of the cable to the breakout board.

attaching the Digital Microwave Sensor

After you've built the structure to mount the Digital Microwave Sensor, attach the sensor to the LEGO structure, as shown in Figure 7-1, with M3 screws through the front of the circuit board. Place nuts on the other side of the structure, as shown in Figure 7-4.

Tighten the screws just enough to hold the sensor in place. Don't overtighten them, since this could crack the sensor's circuit board.

connecting the connector cable and breakout board

Next, you'll need to connect the Digital Microwave Sensor to the EV3 Intelligent Brick. Specifically, you'll build a connection adapter that goes between the LEGO 55805 connector cable and the three-wire cable that came with the Digital Microwave Sensor. You can make this connection by using a breakout board, such as the Breadboard Connector Kit used in Chapter 6.

FIGURE 7-4: Place M3 nuts on the rear of the structure to secure the sensor.

stripping the three-wire cable and jumper wire leads

First prepare the three-wire cable by cutting off the black end of the cable, as shown in Figure 7-5.

Then expose the end of each of the three wires with wire strippers, as shown in Figure 7-6.

You also need to strip both ends of your jumper wire.

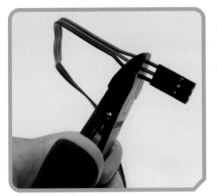

FIGURE 7-5: Cut the black connector off the end of the cable that came with the Digital Microwave Sensor.

FIGURE 7-6: Strip off the ends of all three wires of the cable.

soldering the wires to the breakout board

With your wires ready, solder them as follows to the break-out board, as shown in Figure 7-7:

Solder joint 1 The Digital Microwave Sensor cable's green wire to AN IN

Solder joint 2 One end of the jumper wire to GRND (the one next to AN IN)

Solder joint 3 Both the Digital Microwave Sensor cable's black wire and other end of jumper wire to the GRND next to VBUS

Solder joint 4 The Digital Microwave Sensor cable's red wire to VBUS

These solder connections bring power to the Digital Microwave Sensor on the VBUS (red wire) and GRND (black wire) connections. The signal reading from the Digital Microwave Sensor comes in on the AN IN (green wire) connection. This signal voltage also needs a ground reference, so you put a jumper wire between the two GRND solder connections.

connecting the sensor to the EV3 Intelligent Brick

Next, make the following two connections between the sensor and the EV3 Intelligent Brick:

* Attach a LEGO 55805 connector cable between port 1 of the EV3 Intelligent Brick and the breakout board.

* Attach the white end of the Digital Microwave Sensor's cable into the white connector on the back side of its circuit board.

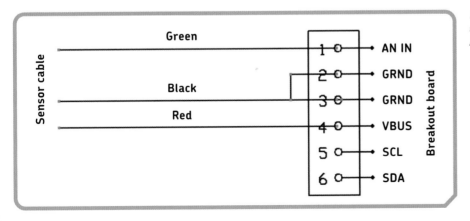

FIGURE 7-7: Diagram for connecting the Digital Microwave Sensor to the breakout board

writing the software

Figure 7-8 shows the program to operate the Motion-Sensing Radar. The software should read the signal from the Digital Microwave Sensor and report an alarm condition when it detects motion.

The program reads the signal from the Digital Microwave Sensor on the green wire of the sensor's cable, which drops to zero voltage when motion is detected. Otherwise, the voltage on this wire should remain at 5V. Follow these steps to write the program:

1. **Loop block** This continuously monitors the output from the sensor and triggers an indication if motion is detected.

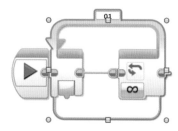

Drag a Loop block onto the screen to place after the Start block.

2. **Raw Sensor Value block** This takes a measurement from the sensor connected to port 1— in this case, the Digital Microwave Sensor. The measurement provided from this block is a value from 0 to 4095, with 0 indicating a zero voltage input and 4095 indicating approximately 5V on the sensor input. The reason the range of values is 0 to 4095 (instead of a nice round 4000 or 5000) is that computers represent numbers using powers of 2, and 4096 is a power of 2 (the count starts at 0, so the maximum value is then 4095).

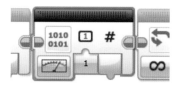

Select a Raw Sensor Value block from the Advanced tab and place it inside the Loop. Set the **Port** menu option to **1**.

3. **Display block** This block prints a label on the front panel of the EV3 Intelligent Brick to remind you of what is being displayed.

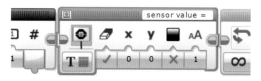

Place a Display block after the Raw Sensor Value block and set the menu options for **Text** to **sensor value =**, **Display** to **Text–Pixels**, **Clear Screen** to **True**, **X** to **0**, **Y** to **0**, **Color** to **False**, and **Font** to **1**.

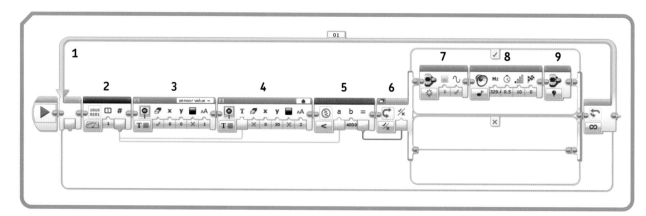

FIGURE 7-8: *The program for the Motion-Sensing Radar reads raw sensor values from port 1 of the EV3 Intelligent Brick.*

4. **Display block**   This shows the value coming from the Digital Microwave Sensor onto the front panel of the EV3 Intelligent Brick. This value will appear below the label you programmed in the previous step.

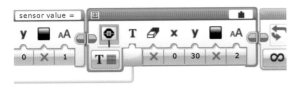

Insert a new Display block after the Display block of the previous step. Set the menu options for **Text** to **Wired**, **Display** to **Text–Pixels**, **Clear Screen** to **False**, **X** to **0**, **Y** to **30**, **Color** to **False**, and **Font** to **2**. Run a wire connection between the output of the Raw Sensor Value block and the input of this Display block.

5. **Compare block**   This checks whether the signal from the sensor has dropped, indicating that the sensor detects motion.

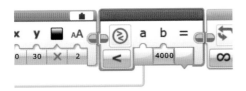

From the Data Operations tab, select a Compare block and place it after the Display block. Set menu options for **Compare** to **Less Than** and set **B** to **4000**. Run a wire from the Raw Sensor Value block output to **A**.

6. **Switch block**   This presents two possible paths based on the result of the Compare block. If the Compare block goes to a true condition, then motion has been detected and the true path of the Switch is taken. Otherwise, the false path is taken.

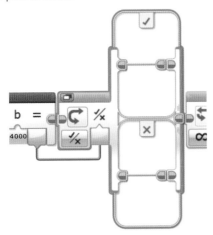

Place the Switch block after the Compare block and set **Switch** to **Logic**. Connect a wire from the Compare block's output to the Switch block's input.

7. **Brick Status Light block**   This lights up the front panel of the EV3 Intelligent Brick when motion is detected.

Place the Brick Status Light block in the true path of the Switch block and set **Brick Status Light** to **On**, **Color** to **1**, and **Pulse** to **True**.

8. **Sound block**   This plays a tone to serve as an alert that motion has been detected. This sound accompanies the flashing light on the front panel of the EV3 Intelligent Brick.

Place a Sound block after the Brick Status Light block and set **Sound** to **Play Tone**, **Frequency** to **329.628**, **Duration** to **0.5**, **Volume** to **10**, and **Play Type** to **0**.

9. **Brick Status Light block**   This shuts off the indicator light on the front panel of the EV3 Intelligent Brick. Because the alarm condition is over, the blinking light is stopped.

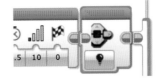

Place a Brick Status Light block at the end of the true path of the Switch. Set **Brick Status Light** to **Off**.

using the motion-sensing radar

After you've completed the program for the Motion-Sensing Radar, activate and test it by waving your hand in front of the Digital Microwave Sensor. The EV3 Intelligent Brick should trigger the light and sound alarm on its front panel. In addition, the yellow LED on the Digital Microwave Sensor should light up when motion is detected.

Test the sensor's range by walking away from the Motion-Sensing Radar. It should detect your movement from many meters away. The Motion-Sensing Radar can also see through doors and walls, so you can aim it through a closed door and know when someone approaches. (See a video of this at my High-Tech LEGO blog site, *http://hightechlego.com/*.) Sensitivity goes down, though, when the microwave radar beam passes through a door or wall. Also, note that the Motion-Sensing Radar can't go through metal and that some older doors are built with layers of steel or aluminum.

You can adjust the sensitivity of the Motion-Sensing Radar by using the blue square knob at the bottom right of the sensor (see Figure 7-2). Use a screwdriver to turn this knob for more or less sensitivity. If you want to detect targets from as far away as possible, raise the sensitivity by turning the knob all the way counterclockwise. The more sensitive you make the sensor, the more false alarms you may get.

In Figure 7-9, the Motion-Sensing Radar is set up to detect motion through a glass window. This allows the Motion-Sensing Radar to stay indoors, where it won't be subject to weather, while still detecting motion outside. To view the EV3 Intelligent Brick from the indoors while keeping the Digital Microwave Sensor pointing outdoors, reverse the LEGO mounting structure's orientation. You should be able to measure motion as far away as 1 meter from the window.

ideas for going further

Instead of sounding an alarm, you can use the Motion-Sensing Radar's motion sensor to trigger another action, such as taking a picture, as you did in Chapter 1 with the Motion-Activated Critter Cam. Combining sensors—say, the LEGO Ultrasonic Sensor with the Motion-Sensing Radar—could be a good way to reduce unwanted alarm triggers. For example, with this sensor combination, you could detect targets that are both moving and within a certain distance.

You could also use the same breakout board configuration and similar software program steps to work with other types of sensors. Any sensor that produces a voltage output of 0V to 5V can work with the setup used in this chapter. Find examples of sensors for detecting force and the concentration of flammable gases in videos at *http://hightechlego.com/*.

what you learned

In this chapter, you built a LEGO-based Doppler radar for detecting objects in motion. This motion detector can serve as an alarm system that senses movement over a distance of many meters and even on the other side of obstructions such as walls, doors, or glass. You saw how to read voltage signals by using MINDSTORMS EV3, which is compatible with many other possible sensors. To send the sensor signals to the EV3 Intelligent Brick, you built an adapter that connects to EV3 cables.

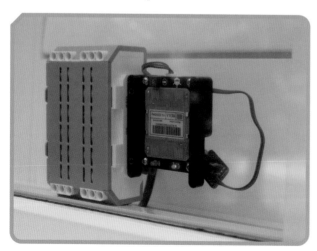

FIGURE 7-9: *The Motion-Sensing Radar detects motion outside, through a window.*

8

the tower of Eratosthenes

Contrary to popular belief, no explorer or scientist in the days of Christopher Columbus thought the earth was flat. In fact, people in ancient times knew that the earth was round from watching ships come over the horizon, studying the motion of stars, observing lunar and solar eclipses, and contemplating shadows cast by the sun.

Ancient people not only knew the shape of the earth but also could measure the diameter and axial tilt of the earth with surprising accuracy. Eratosthenes, born in 276 BCE, discovered a way to measure the circumference of the earth. In this chapter, you'll use LEGO parts and some simple tools to re-create the tower Eratosthenes used to make his measurement. Then you'll learn how to calculate the latitude and tilt of the earth by measuring the angle cast by the sun's shadow at different times of the year.

Keep in mind that you'll have to take measurements several months apart, so after building the device and making a first measurement, set the device aside for a few months to use again.

understanding Eratosthenes's discovery

The technique for measuring the circumference of the earth occurred to Eratosthenes after hearing about a well in the town of Swenet, some 900 km directly south from his home in Alexandria. At noon every year on the summer solstice, the sun illuminated the water at the bottom of the well, but not the sides of the well, meaning that the sun was directly overhead.

Eratosthenes realized that if he measured the angle of the sun on a solstice day in Alexandria at the same time the sun was directly overhead in Swenet, he could calculate the circumference of the earth. The geometry of this idea is shown in Figure 8-1. The only tool Eratosthenes needed for this measurement was a pole held perpendicular to the ground so that he could measure the angle of the pole's shadow.

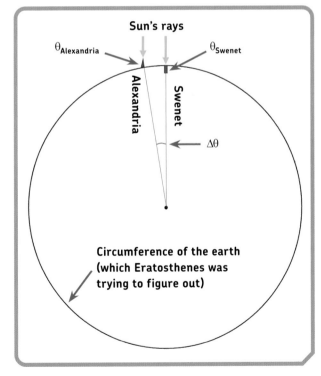

FIGURE 8-1: Eratosthenes used the angle of the sun on the summer solstice to calculate the circumference of the earth.

Figure 8-1 shows several important angles:

$\theta_{\textbf{Alexandria}}$ The angle of the shadow cast by the pole situated at Alexandria. This angle was measured by Eratosthenes.

$\theta_{\textbf{Swenet}}$ The angle of the sun's rays down a well in the town of Swenet. Eratosthenes knew that this angle was zero at noon on a particular day. An observer knows that the angle is zero by seeing that the bottom of the well is brightly lit by sunlight but the sides of the well are not lit.

$\Delta\theta$ The difference in angle between $\theta_{Alexandria}$ and θ_{Swenet}. We can use the $\Delta\theta$ angle to calculate the circumference of the earth if the distance between the two towns is also known. Eratosthenes knew the distance between the two towns by talking with people who often walked between them.

We can use shadow angles to figure out other measurements, too, such as our latitude and the tilt of the earth. Unlike Eratosthenes's experiment, which had to take place at two far-apart locations, we can take the shadow angle measurements at one location.

building the tower of Eratosthenes

To re-create Eratosthenes's experiments, you'll build a Tower of Eratosthenes out of LEGO pieces (Figure 8-2).

The purpose of the tower is to cast a shadow long enough to measure. It should be tall enough that its shadow is long enough to allow for a margin of error in measurement, but not so tall that the tower tends to fall over. To cast a distinct shadow, the top of the tower should be narrow. The tower in Figure 8-2 is 350 mm tall.

what you'll need

Figure 8-3 shows the LEGO parts you need to build the Tower of Eratosthenes. All of the bricks used in this chapter are included in the MINDSTORMS EV3 #31313 set.

You'll need a board, such as a 50 cm by 50 cm plywood square, on which to place the tower and mark shadow position.

FIGURE 8-2: The Tower of Eratosthenes is built from parts of the MINDSTORMS EV3 #31313 set.

You'll also need the following tools:

* Bubble level (you can find this at a hardware store)
* Ruler or tape measure
* Pen or marker
* Pad of paper squares, such as a pad of 7.5 cm (3 in.) square sticky notes

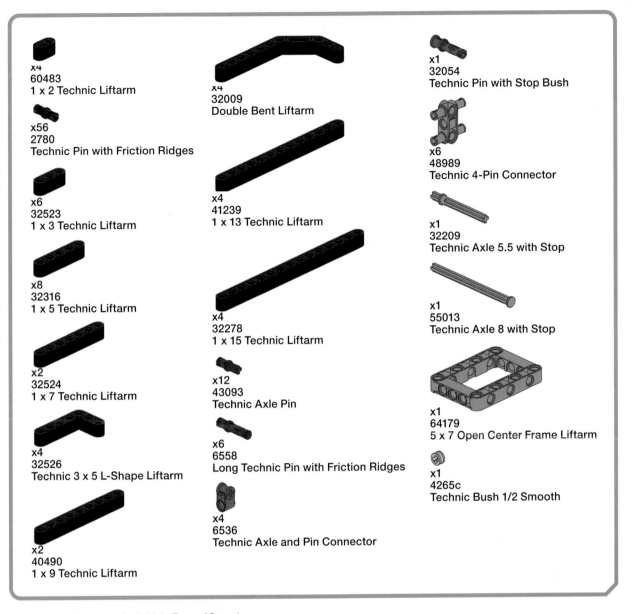

x4
60483
1 x 2 Technic Liftarm

x56
2780
Technic Pin with Friction Ridges

x6
32523
1 x 3 Technic Liftarm

x8
32316
1 x 5 Technic Liftarm

x2
32524
1 x 7 Technic Liftarm

x4
32526
Technic 3 x 5 L-Shape Liftarm

x2
40490
1 x 9 Technic Liftarm

x4
32009
Double Bent Liftarm

x4
41239
1 x 13 Technic Liftarm

x4
32278
1 x 15 Technic Liftarm

x12
43093
Technic Axle Pin

x6
6558
Long Technic Pin with Friction Ridges

x4
6536
Technic Axle and Pin Connector

x1
32054
Technic Pin with Stop Bush

x6
48989
Technic 4-Pin Connector

x1
32209
Technic Axle 5.5 with Stop

x1
55013
Technic Axle 8 with Stop

x1
64179
5 x 7 Open Center Frame Liftarm

x1
4265c
Technic Bush 1/2 Smooth

FIGURE 8-3: LEGO parts used to build the Tower of Eratosthenes

constructing the tower

You'll build the Tower of Eratosthenes in two sections: the base and the top. The base has a couple of features to make taking measurements a bit easier, as you'll do later in this chapter. The top can be built in various ways, so you can get creative, but the instructions in this chapter show one particular build.

building the base

A platform in the base of the tower, created with an open center frame liftarm, provides a surface on which to place a bubble level, the circular device in Figure 8-4.

FIGURE 8-4: *A bubble level placed inside the base of the Tower of Eratosthenes indicates whether the tower is level.*

In this section, you'll build that base, including two features that will help you use the Tower of Eratosthenes. First, a Technic axle on the underside of the structure will indicate the center of the tower, allowing you to precisely measure from that point. Second, the level will check whether the tower is perpendicular to the ground, giving you a more accurate measurement.

Follow the steps shown here to build the tower base.

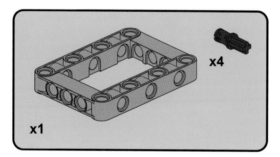

1

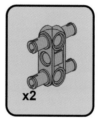

2

3

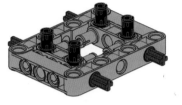

4

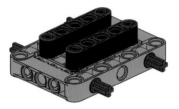

5

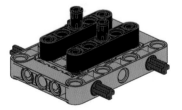

6

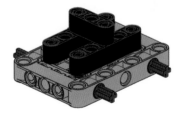

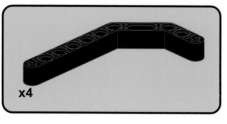

x4

7

x8

8

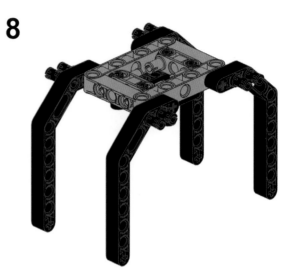

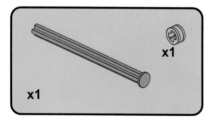

x1

x1

9

x1

10

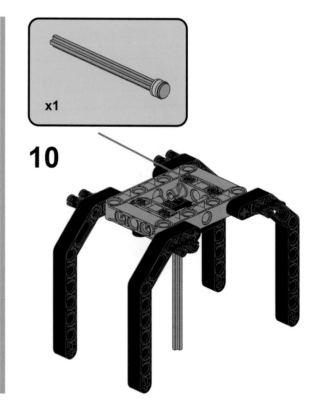

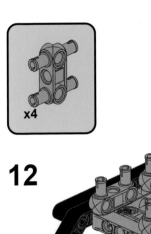

11

x2

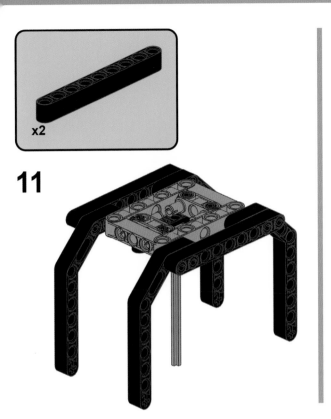

12

x4

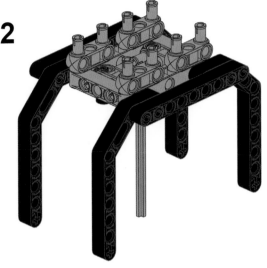

13

x8

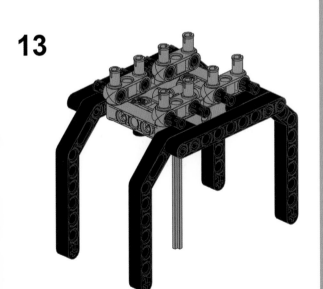

14

x4

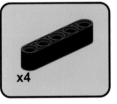

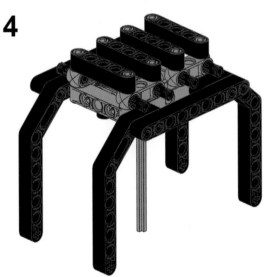

building the top of the tower

You can build the upper portion of the tower by attaching liftarms to the Technic pins placed in step 13 of the building instructions. The following instructions show one way to attach a tower to the base, but there are many ways you can build it.

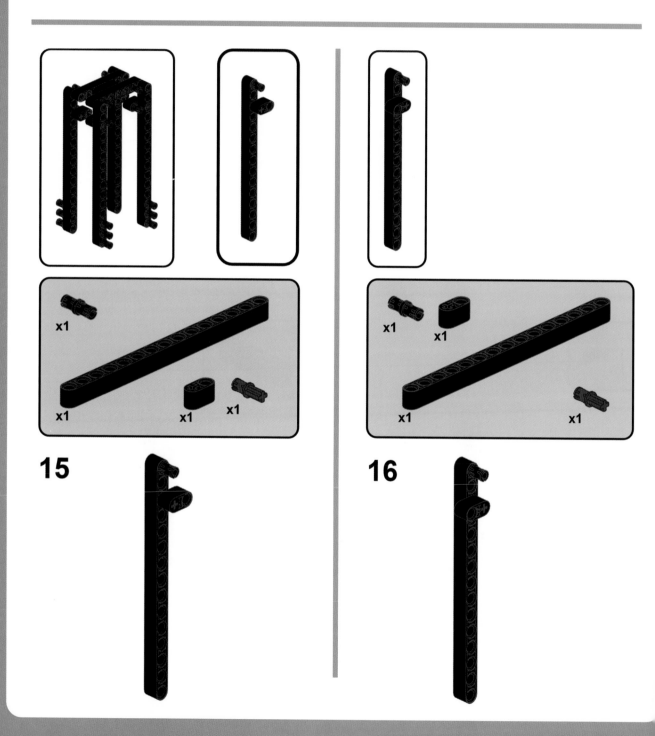

17

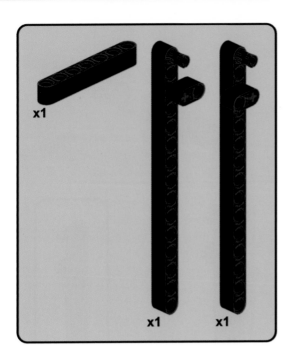

x1

x1 x1

18

x3

19

x1

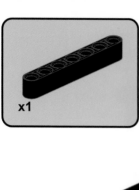

x1

20

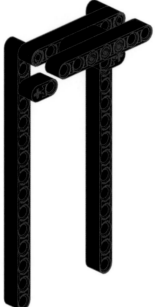

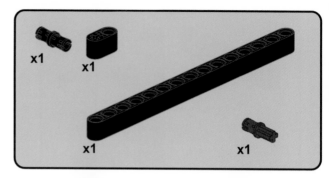

x1   x1

x1   x1

21

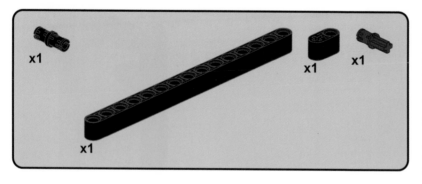

x1

x1   x1

x1

22

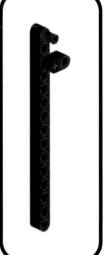

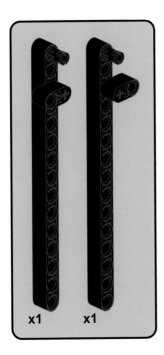

x1 x1

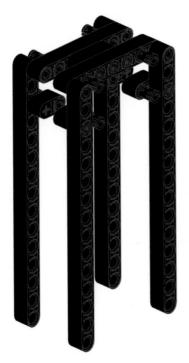

23

x8

24

x12

25

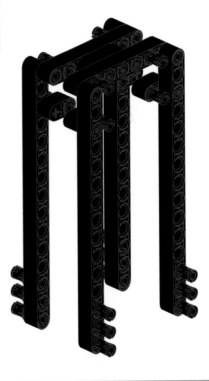

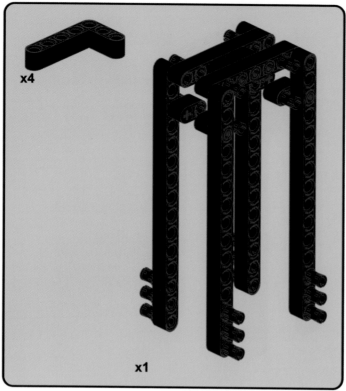

x4

x1

26

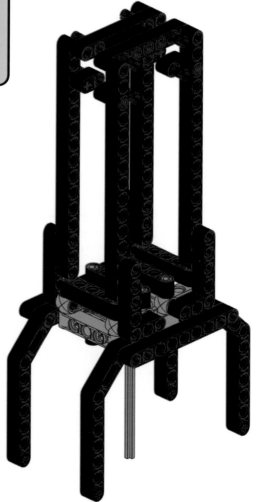

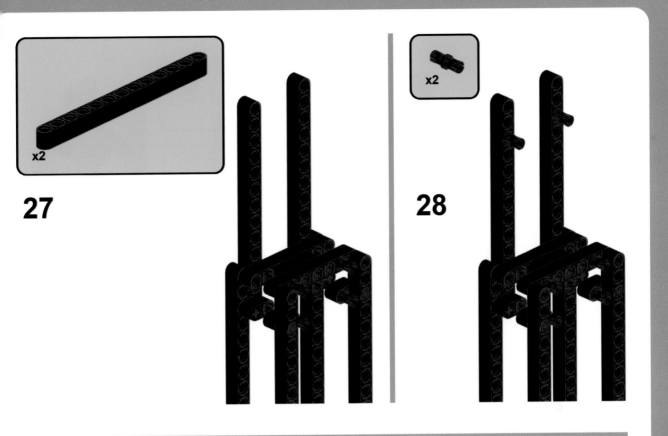

27

x2

28

x2

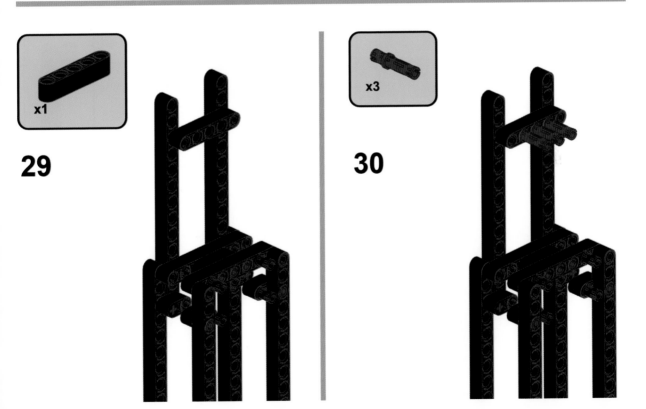

29

x1

30

x3

31

x1 x2

32

x1 x2

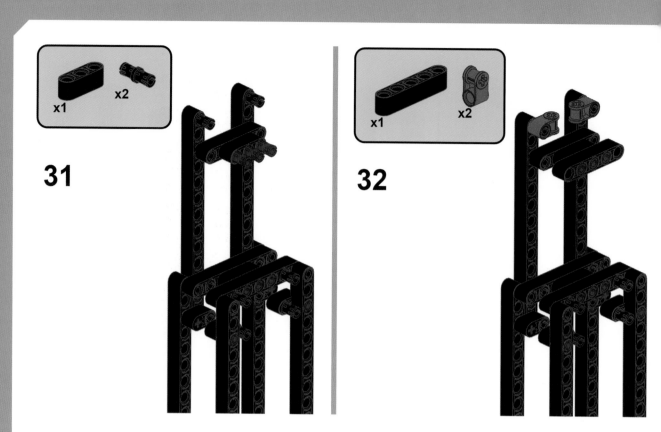

33

x2

34

x1 x1 x1

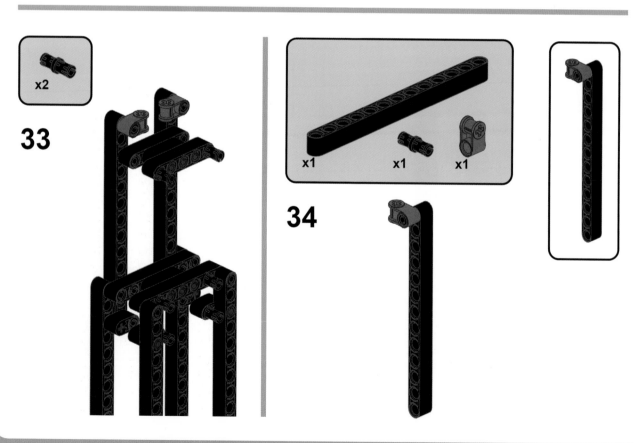

35

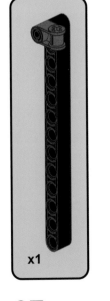

x1

36

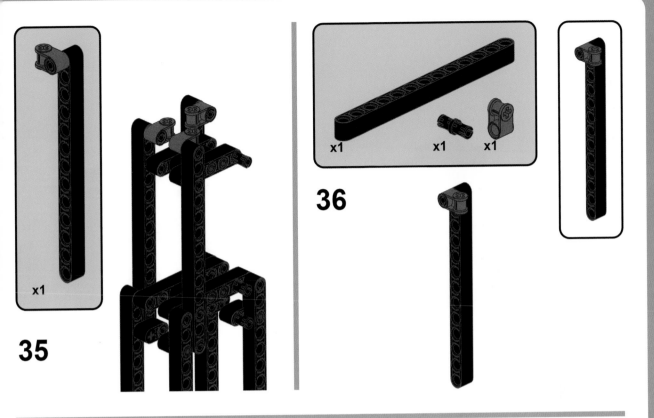

x1   x1   x1

37

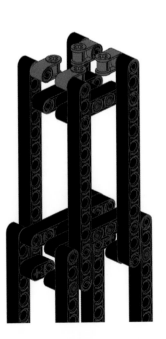

x1

38

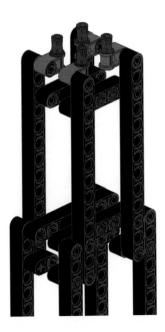

x4

x2

39

x2

40

x1

41

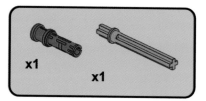

x1 x1

42

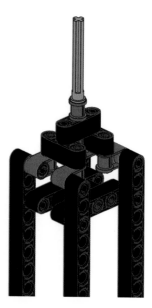

leveling the base

Place the bubble level on the base, as shown in Figure 8-5. Place slips of paper under the legs of the tower until the bubble is centered within the rings printed on the level. In the example shown in Figure 8-5, the bubble level indicates that the tower needs adjustment—the bubble is offset low and to the left, so you would add slips of paper to the upper-right tower leg to level the tower.

FIGURE 8-5: *A bubble level placed on the base of the tower indicates when the tower is oriented straight up.*

understanding the angle of the sun

You can use the angle of the sun to determine both the tilt of the earth and your current latitude. To do this, you'll measure the shadows cast at solar noon on specific days of the year. *Solar noon* is the time at which the sun reaches its highest elevation of the day, casting the shortest shadow, and it doesn't always correspond to 12 PM on a clock. To find the clock time of solar noon for a particular day and location, you can go to websites that track the motion of the sun, such as *https://www.timeanddate.com/sun/, https://www.esrl.noaa.gov/gmd/grad/solcalc/,* or *http://suncalc.net/.*

The tilt of the earth relative to the oncoming rays of sunlight changes because of earth's orbit around the sun. But, on several special days of the year, the angle becomes simplified. These special days are the equinoxes and solstices, diagrammed in Figure 8-6 and explained in detail in the following sections.

To conduct this experiment, you'll have to measure the angle of the sun on these specific days.

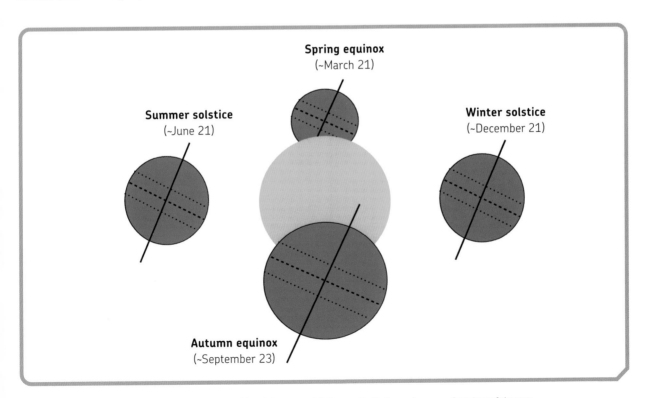

FIGURE 8-6: *As the earth orbits around the sun, the earth's axis has a special alignment with the sun's rays on four days of the year.*

equinoxes

Equinoxes occur on March 19, 20, or 21 and again on September 22, 23, or 24, when the sun's rays are directly perpendicular to the earth's axis of rotation.

On the day of an equinox, the hours of the night and day are almost equal. The tilt of the earth's axis of rotation has no influence on the angle of the sun because the sun's rays are exactly perpendicular to the earth's axis. The angle of the sun at solar noon on the equinoxes is equal to the observer's latitude. In mathematical terms, this is

$$\theta_{equinox} = \theta_{latitude}$$

where $\theta_{equinox}$ is the angle measured for the sun at solar noon on the day of an equinox. Figure 8-7 describes the angles involved. By a principle in geometry known as *equal alternate angles*, $\theta_{latitude}$ is the same as $\theta_{equinox}$.

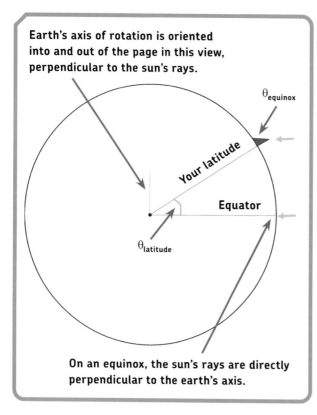

FIGURE 8-7: *The angle of the shadow cast by the Tower of Eratosthenes is equal to your latitude at solar noon on the day of an equinox.*

summer solstice

The *summer solstice* occurs on June 20, 21, or 22 in the Northern Hemisphere and on December 20, 21, 22, or 23 in the Southern Hemisphere. On the summer solstice, the earth's axis of rotation is pointed directly *toward* the sun. Figure 8-8 illustrates the summer solstice for the Northern Hemisphere. Seasons are reversed for the Southern Hemisphere.

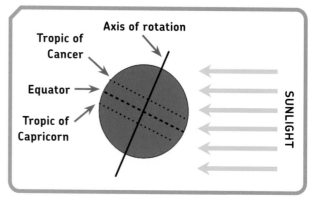

FIGURE 8-8: *The summer solstice occurs when the earth's axis of rotation is tilted directly toward the sun.*

The summer solstice is the longest day of the year, and the sun's path in the sky will reach the highest point for the year. For observers north of the the Tropic of Cancer or south of the Tropic of Capricorn, the angle of the sun at solar noon on the summer solstice is equal to the observer's latitude minus the tilt of the earth.

In mathematical terms, we can express this as

$$\theta_{summer} = \theta_{latitude} - \theta_{tilt}$$

where θ_{summer} is the sun's angle measured on solar noon of the day of the summer solstice, $\theta_{latitude}$ is the angle of the observer's latitude, and θ_{tilt} is the angle of the earth's tilt. These angles are diagrammed in Figure 8-9. By the geometric principle of equal alternate angles, the θ_{summer} angle is the same as the angle measured from the shadow cast by the Tower of Eratosthenes. Thus, a measurement of the sun's angle on a summer solstice will give information about the combination of latitude and tilt.

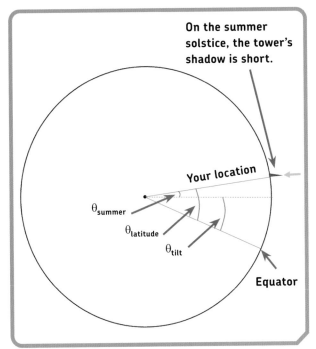

On the summer solstice, the tower's shadow is short.

Your location

θ_{summer}

$\theta_{latitude}$

θ_{tilt}

Equator

FIGURE 8-9: The angle of the shadow cast by the Tower of Eratosthenes at solar noon on the summer solstice is equal to your latitude minus the tilt of the earth.

winter solstice

The *winter solstice* occurs on December 20, 21, 22, or 23 in the Northern Hemisphere and on June 20, 21, or 22 in the Southern Hemisphere. During the winter solstice, the earth's axis of rotation is pointed directly *away* from the sun, as shown in Figure 8-10.

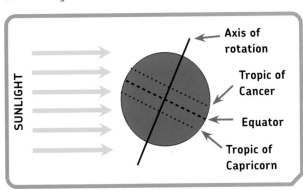

SUNLIGHT

Axis of rotation

Tropic of Cancer

Equator

Tropic of Capricorn

FIGURE 8-10: The winter solstice occurs half a year after the summer solstice, when the earth's axis of rotation is tilted directly away from the sun.

The winter solstice is the shortest day of the year, when the sun travels across the lowest path through the sky. For observers north of the Tropic of Cancer or south of the Tropic of Capricorn, the angle of the sun at solar noon is equal to the observer's latitude plus the tilt of the earth.

In mathematical terms, we can express this as:

$$\theta_{winter} = \theta_{latitude} + \theta_{tilt}$$

where θ_{winter} is the angle measured for the sun at solar noon on the day of the winter solstice, $\theta_{latitude}$ is the angle of the observer's latitude, and θ_{tilt} is the angle of the earth's tilt. These angles are diagrammed in Figure 8-11. By the geometric principle of equal alternate angles, the θ_{winter} angle is the same as the angle measured from the shadow cast by the Tower of Eratosthenes. Thus, a measurement of the sun's angle on a winter solstice will give information about the combination of latitude and tilt.

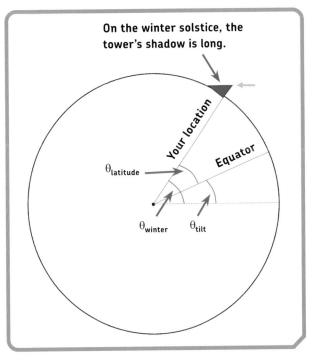

On the winter solstice, the tower's shadow is long.

Your location

Equator

$\theta_{latitude}$

θ_{winter}

θ_{tilt}

FIGURE 8-11: The angle of the shadow cast by the Tower of Eratosthenes at solar noon on the winter solstice is equal to your latitude plus the tilt of the earth.

measuring with the tower of Eratosthenes

To figure out the tilt of the earth and your current latitude, you have to measure the angle of the sun at solar noon on an equinox and then measure it again on a solstice. The shadow cast by the Tower of Eratosthenes will have different lengths on an equinox, winter solstice, and summer solstice, as shown in Figure 8-12.

The angle on the summer solstice (the green dashed curve) is the smallest angle of the three, followed by an equinox's angle (the dashed purple curve) and the winter solstice angle (the dashed orange curve).

Since you want to figure out two pieces of information—your latitude and the tilt of the earth—you need two different measurements to work with. That's because, if you have two variables to figure out, you need at least two equations involving these variables.

With this information, it's time to take the Tower of Eratosthenes outside!

measuring on a solstice

On either the summer or winter solstice, place the tower on a board, which serves as a stable surface that you can mark with a pen. The sun will make a shadow of the tower, such as the one shown in Figure 8-13, a photo taken on a winter solstice.

Use a pen to make a mark on one end of the wood board and then position the center of the tower over this mark. The axle pointer in the base of the tower is useful for finding the center of the tower—place the tip of the axle on the mark. While keeping the center mark aligned, level the tower by using slips of paper and the bubble level.

At solar noon, mark the tip of the tower's shadow on the board with a dot.

Now you can calculate the angle of the sun by using trigonometry. First, use a tape measure, ruler, or yardstick to measure the distance from the axle under the tower to the shadow tip you just marked. It's okay to move the tower out of the way and off the board, since you've already made the dots marking the shadow length. In the case of Figure 8-13, the shadow's length measured 626 mm.

FIGURE 8-12: Angles of the sun at different times of the year

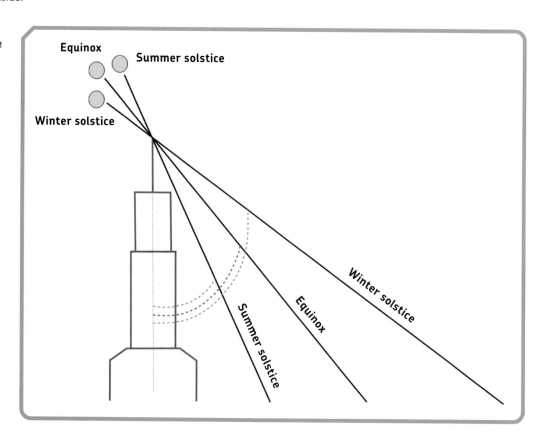

FIGURE 8-13: *The Tower of Eratosthenes is set on a board with the center marker placed over a dot drawn onto the board.*

Figure 8-14 illustrates how to calculate the angle of the sun by using trigonometry.

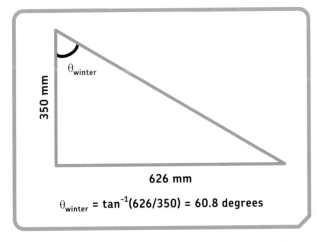

$$\theta_{winter} = \tan^{-1}(626/350) = 60.8 \text{ degrees}$$

FIGURE 8-14: *Calculating the angle of the sun, given the height of the Tower of Eratosthenes (350 mm)*

If we know that the shadow length at solar noon on the winter solstice is 626 mm and that the tower is 350 mm tall, using an inverse tangent calculation, we can calculate that the angle of sun on the winter solstice, θ_{winter}, is 60.8 degrees, because we know that $\tan(\theta_{winter})$ must equal 626 mm / 350 mm.

Because the measurement in this example was taken on the winter solstice, this angle is equal to the tilt of the earth plus the latitude. Now we need another measurement taken on an equinox to provide more data.

TANGENT

Tangent is a function in trigonometry relating an angle in a right triangle to the length of the sides of the triangle. As in Figure 8-15, the sides of the right triangle are called *opposite*, *adjacent*, and *hypotenuse* in relation to angle A. The tangent of angle A in equation form is as follows:

$$\tan(A) = \text{opposite/adjacent}$$

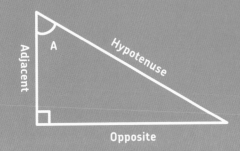

FIGURE 8-15: *The tangent function is a relationship of an angle to the length of the sides of a right triangle.*

measuring at an equinox and finding the latitude

On an equinox, make an angle measurement following the same procedure as the one explained in "Measuring on a Solstice" on page 108. On a recent spring equinox at my location, the length of the shadow was 271 mm at solar noon.

To find my geographic latitude, I performed trigonometry with the measured shadow length:

$$\theta_{equinox} = \tan^{-1}(271/350) = 37.7 \text{ degrees}$$

Because the earth's axis on an equinox is perpendicular to the rays of the sun, the tilt of the earth does not affect the angle measured, leaving only the observer's latitude. You can always check your latitude calculation by using Google Maps or other sources. For example, the actual latitude of my location was 37.1 degrees, so the measurement I made had a 0.6 degree error. This error can come from failing to perfectly level the Tower of Eratosthenes, making the measurement on a day other than the equinox (for reasons such as weather), or lacking precision when measuring the shadow with a tape measure.

finding the tilt of the earth

The calculation you made from measurements taken on an equinox was simple, since the angle measured involved only your latitude and not the tilt of the earth's axis. But the tilt of the earth's axis does come into play on a solstice.

Therefore, there are two variables in the shadow angle you measure on a solstice—the tilt and the latitude—but since you've now figured out your latitude, you can use this information to determine the tilt of the earth. For example, with two measurements made on the winter solstice and spring equinox, you can rearrange the equation on page 107 as follows:

$$\theta_{tilt} = \theta_{winter} - \theta_{equinox}$$
$$= 60.8 - 37.7 \text{ degrees}$$
$$= 23.1 \text{ degrees}$$

The actual value of θ_{tilt} is 23.5 degrees, so the measured tilt in this example has an error of 0.4 degrees.

This chapter showed data captured on a winter solstice combined with an equinox, but you could also combine a summer solstice with an equinox. In that case, you would solve for the following equation:

$$\theta_{tilt} = \theta_{equinox} - \theta_{summer}$$

ideas for going further

If you wanted to re-create Eratosthenes's experiment for determining the circumference of the earth, you could measure the latitude at two locations that are at different latitudes, meaning they're separated in the north-south direction. This could make a good joint project for students at two schools located in different parts of the country—for example, in Miami and New York City—or even two different countries.

The two locations have to be separated by more than a few hundred kilometers in the north-south direction for the calculation to work well. The distance in the east-west direction doesn't matter.

Make your latitude measurements at your two locations and then use the following equation:

$$\text{earth circumference} = 360/\Delta\theta \cdot \Delta x$$

Here, $\Delta\theta$ is the difference in latitude between the two observers, and Δx is the distance between the two observers in the north-south direction. In Figure 8-1 on page 89, $\Delta\theta$ is the angle between Swenet and Alexandria, and Δx is the distance between Swenet and Alexandria.

what you learned

In this chapter, you measured the latitude at your location and the tilt of the earth's axis. You could look up either of these values on a smartphone, but now you know how to make the measurements for yourself. In addition, you've learned the astronomical significance of solar noon, solstices, and equinoxes. You've also learned the trigonometry to convert the measured length of a shadow to a solar angle, which, in turn, gives you the latitude and earth's tilt angle. LEGO parts provided a convenient way to build a vertical tower for casting a shadow, including a means to mark both ends of the shadow and make the tower level.

hacking LEGO light bricks

The LEGO light brick is useful for all sorts of light-based designs, but it's available in only two colors: red or yellow. In this chapter, you'll hack the light brick, shown in Figure 9-1, to make it any color you want. You can even make it flash, change colors, or emit infrared and ultraviolet (UV) light.

FIGURE 9-1: *The LEGO light brick has an LED embedded inside a brick.*

Then you'll build a device that uses the light brick to project a beam, simulating a laser, by using a lens to shape the light's output.

what you'll need

Figure 9-2 shows the LEGO parts used in this chapter. Parts not included in the MINDSTORMS EV3 #31313 set are shown with an asterisk.

Light bricks come in a case that is either red (part number 54930c01) or clear (part number 54930c02). They're otherwise identical, so you can select whichever case you prefer for the projects in this chapter.

In addition to these LEGO parts, you'll need the following:

* A 3 mm LED in the color of your choice. LEDs are available at online stores, including Lighthouse LEDs (*https://lighthouseleds.com/*), Digi-Key Electronics (*https://digikey.com/*), Mouser Electronics (*https://mouser.com/*), and Amazon (*https://www.amazon.com/*). You can also find kits containing many different color selections.

* #1 Phillips head screwdriver

* Tweezers

* Diagonal cutters

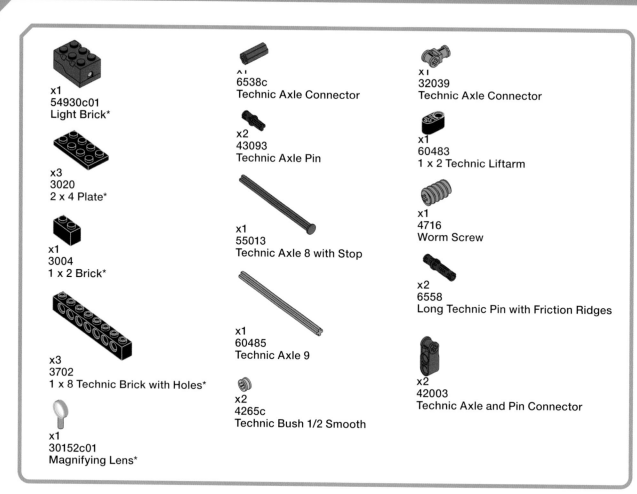

x1
54930c01
Light Brick*

x3
3020
2 x 4 Plate*

x1
3004
1 x 2 Brick*

x3
3702
1 x 8 Technic Brick with Holes*

x1
30152c01
Magnifying Lens*

^1
6538c
Technic Axle Connector

x2
43093
Technic Axle Pin

x1
55013
Technic Axle 8 with Stop

x1
60485
Technic Axle 9

x2
4265c
Technic Bush 1/2 Smooth

x1
32039
Technic Axle Connector

x1
60483
1 x 2 Technic Liftarm

x1
4716
Worm Screw

x2
6558
Long Technic Pin with Friction Ridges

x2
42003
Technic Axle and Pin Connector

FIGURE 9-2: LEGO parts used to hack the light brick

changing the colors of LEGO light bricks

You can change your light brick to be any color you want. For example, the bricks on either end of Figure 9-3 switch between red, green, and blue in both slow and fast cycles. (Watch videos of these color-changing bricks at *http://hightechlego.com/*.) The other three bricks are pink, green, and blue.

The fun of this project is in picking out the colors you want to try, so search carefully! Whatever color you choose, make sure to get the size called *3 mm* (or sometimes *T1*), since this is the size that fits in the light brick.

FIGURE 9-3: These five LEGO light bricks have been modified for customized colors.

unscrewing the light brick

First, remove the screw on the base of the light brick. Use a #1 Phillips head screwdriver to remove the screw, as shown in Figure 9-4.

FIGURE 9-4: Use a #1 Phillips head screwdriver to remove the base of the light brick.

You should now see the guts of the light brick (Figure 9-5). Make a note of how the batteries are oriented, since the direction in which you insert them matters.

FIGURE 9-5: Removing the cover of the light brick shows its internal components.

Before you can remove the factory-installed LED, you have to remove the batteries, which you can do with tweezers. Then use the tweezers to pluck out the LED. Notice that the LED leads are bent, as shown in Figure 9-6.

You'll have to re-create this orientation in the new LED.

FIGURE 9-6: Remove the batteries and factory-installed LED from the light brick.

reshaping the leads

Next, bend the leads of your new LED into the same shape as the leads of the factory LED you just removed. Figure 9-7 shows six steps to shape your replacement LED.

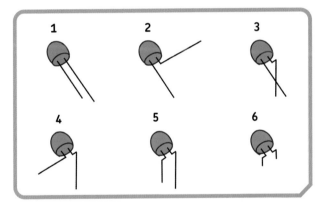

FIGURE 9-7: Bend and trim the new LED to fit into the LEGO light brick with these six steps.

1. Hold the new LED so that the longer lead is on the right.

2. Bend the longer lead sideways at a right angle. Make the bend 2 to 3 mm away from the base of the LED.

3. Bend the longer lead again at another right angle so that the lead is pointing downward. Make this downward bend 1 to 2 mm away from the bend created in step 2.

4. Bend the shorter lead (the lead on the left) sideways at a right angle, away from the longer lead. Make the bend 2 to 3 mm away from the base of the LED.

5. Bend the shorter lead at a right angle again, such that the lead is pointing downward. Make this downward bend 1 to 2 mm away from the bend created in step 4.
6. Use diagonal cutters to trim both leads to a length of 3 to 4 mm below the bend made in step 5.

The new LED should look like the examples in Figure 9-8. Now that you've shaped your new LED, place it into the light brick. Then put the batteries back in, taking care to get the battery polarity correct, and screw the base back on. Your custom-color light brick is ready for action. Test it by holding down the switch on the back of the brick.

FIGURE 9-9: *You can see an infrared LED installed in a light brick if you view it through a night vision device, as described in Chapter 5.*

FIGURE 9-8: *Three customized LEDs are shown here after the bending/trimming procedure.*

performing infrared and UV light brick experiments

Your custom-colored light brick can decorate pretty much any LEGO model, but you can also use it for scientific purposes. For example, try installing an infrared or UV LED in a light brick. The infrared LED can serve as a covert source of illumination, along the lines of the project built in Chapter 5. In Figure 9-9, you can see such a brick through a night vision monocular.

We used UV light in Chapter 6 to attract insects. The project required a very bright UV source, much stronger than a small battery-powered LED can provide. But sometimes you might want a more compact and less powerful UV light—for instance, when you're working with *fluorescence*, a phenomenon in which visible light is emitted from a substance when illuminated with UV light. Substances that fluoresce include the ink used in highlighter pens; pigments in the skin of some fish; and some minerals, including the gemstones ruby, emerald, and diamond.

One of the many applications of fluorescence is as an anticounterfeiting measure for money. Many countries embed a strip into currency notes that glows in ultraviolet light, serving as a way to check that the currency is genuine. For example, Figure 9-10 shows a $50 bill illuminated by a UV light brick. Notice that the embedded strip glows yellow.

FIGURE 9-10: *An ultraviolet LED installed in a light brick can serve as a tool to check for counterfeit money. The glowing yellow strip on this $50 bill verifies that it's genuine.*

All US currency notes of $5 and higher have such embedded security strips, with different fluorescent colors for each denomination. Euro currency notes have fluorescent fibers embedded in the paper, along with text and graphical features that become visible under ultraviolet light. The specific LED used for the example in Figure 9-10 is available from Lighthouse LEDs (part number 3MMROUNDTOPLEDUVPURPLE).

building a light brick that remains on

In some experiments, you may want to keep your light brick turned on rather than having to continuously hold down the button on the back of the light brick. The following building instructions use a 1 × 2 brick to keep the button pressed. When you're not using the light brick, just remove this 1 × 2 brick.

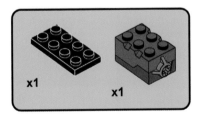

1

2

creating a light beam

The output from an LED has a wide *divergence*, which means it projects a beam of light with a wide angle. For some purposes, you may want to gather this spread of light into as narrow a beam as possible, a technique known as *collimation*. The collimated beam could replace a laser in situations that call for a dim beam, such as in a model or display in which a laser could pose a safety concern. To accomplish collimation, you could place a lens in front of the light brick. Conveniently, LEGO makes its own: the magnifying lens, most commonly used in accessories for minifigures, such as in the detective in Figure 9-11.

FIGURE 9-11: *The LEGO minifigure accessory magnifying glass is actually a functioning lens.*

We'll use the magnifying lens to turn the light brick's output into a beam.

calculating the magnifying glass's focal length

Lenses are curved pieces of glass or transparent plastic shaped to bend light rays. As diagrammed in Figure 9-12, all lenses have a *focal length*, which is the distance from the lens at which light rays hitting it from a distant object come to a point.

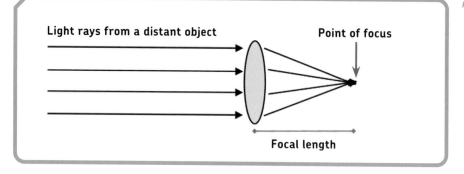

FIGURE 9-12: *The focal length of a lens*

Light rays from a distant object

Point of focus

Focal length

LEGO hasn't published specifications on the focal length of the magnifying lens. However, you can measure it by holding the lens horizontally under an overhead light fixture. The focal length is the distance from the lens at which the elements of the light fixture form a sharp image, giving you a nonblurry view of the features in the light fixture.

Move the lens up and down until the elements of the light fixture come into focus on a piece of paper. In Figure 9-13, the overhead light is a desk lamp that uses a circular array of bulbs as a light source.

FIGURE 9-13: *You can find the focal length of a lens by holding the lens underneath a light fixture and looking for the elements of the light fixture to come into focus.*

This circular array becomes clear at the distance under the lens corresponding to the lens's focal length. In this experiment, this distance was 20 mm.

building a worm drive to align the lens

To collimate the beam from a light brick, you have to place a lens at a distance of one focal length from the light brick, producing an effect that is the reverse of the one from Figure 9-12: light rays that come from a small light source are bent by the lens to travel in a straight line. For this to work, you must align the lens with the light brick and make sure it has the same angle as the light brick's face. You can accomplish this alignment by placing the lens in the hand of a minifigure, as shown in Figure 9-14. But this process isn't very precise, and it's easy to knock the lens out of alignment.

FIGURE 9-14: *A minifigure can serve as a mount to position a lens for collimation.*

An alternate technique is to build a mechanism that holds and slides the lens in front of the light brick, as pictured in Figure 9-15. This mechanism uses a *worm drive*, which consists of two gears: a *worm screw* (the spiral-shape gear) and a *worm gear* (the Technic bush). Follow these directions to build the worm drive.

FIGURE 9–15: *Turning the red adjustment knob at the back of this worm drive slides the lens into place.*

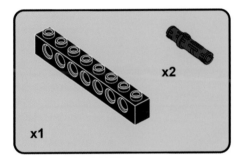

x2

x1

1

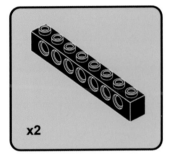

x2

2

x2

3

x2

4

x2

5

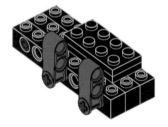

x1

x1

6

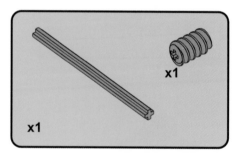

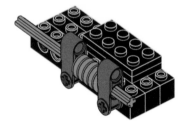

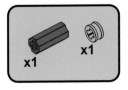

x1 x1

x1

x1

7

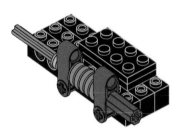

8

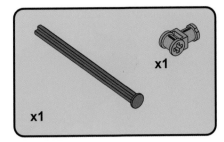

x1

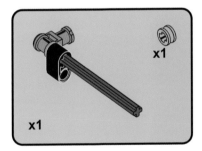

x1

x1

9

10

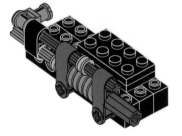

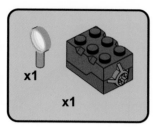

11

testing the light beam

After you've built the worm-drive mechanism, check that the magnifying lens is installed with the lens parallel to the face of the light brick. Also check that the center of the lens lines up with the center of the light brick, which should be the case if the magnifying lens is pressed all the way into the technic axle connector. Then turn on the light brick. Look at the image cast through the lens on a wall or piece of paper about 50 cm away. You'll see the image change shape and size as you adjust the red knob on the back of the worm-drive mechanism. At collimation, an image like Figure 9-16 will appear, with a square beam shape about 2 cm wide.

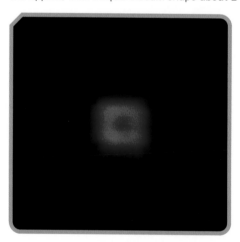

FIGURE 9-16: The collimated beam shape from a light brick is a square, seen here projected on a wall 50 cm away from the light brick.

The beam is square because the tiny area inside the LED where the light is produced is square. If you want a circular, spotlight effect, you can slide the lens a little farther away from the LED.

what you learned

In this chapter, you learned how to modify LEGO light bricks to create lights of whatever color you'd like, rather than having to stick with the two factory options of red and yellow. Use your new color choices in LEGO models to add lighting effects, or modify light bricks with infrared and UV LEDs to use in science projects. You also learned how to work with lenses to change the shape of the light brick's beam. Adding a lens to a light brick can collimate the light beam to create a more directed beam. In the next chapter, you'll explore LEDs further by programming color and brightness that changes.

10

the flickering fireplace

Although hacking a light brick as you did in Chapter 9 can give you a brick with a customized color, it's still a constant color. You may want to light up your LEGO creations with colors or brightness levels that change over time. One way to do this is with a device called the BlinkM, shown in Figure 10-1. The BlinkM is an LED that you can program to have whatever color pattern or brightness you like.

In this chapter, you'll use the BlinkM to create a fireplace for LEGO minifigures to enjoy on a cold evening, as shown in Figure 10-2. The fireplace will change colors (red, orange, yellow, and a dash of blue) as well as flicker in brightness, the way flames do.

FIGURE 10-2: A BlinkM is installed here to build a LEGO fireplace.

You can see a video of the fireplace at *http:// hightechlego.com/*.

FIGURE 10-1: *The BlinkM is a programmable LED.*

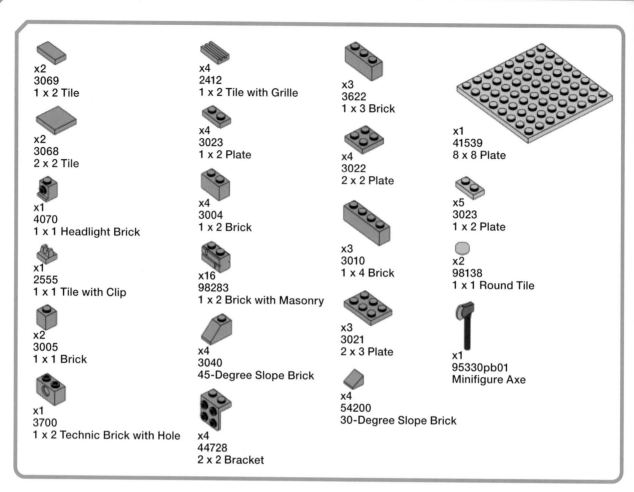

FIGURE 10-3: *LEGO parts for building the BlinkM LEGO fireplace structure*

The following are the LEGO parts shown in Figure 10-3:

- x2 3069 1 x 2 Tile
- x2 3068 2 x 2 Tile
- x1 4070 1 x 1 Headlight Brick
- x1 2555 1 x 1 Tile with Clip
- x2 3005 1 x 1 Brick
- x1 3700 1 x 2 Technic Brick with Hole
- x4 2412 1 x 2 Tile with Grille
- x4 3023 1 x 2 Plate
- x4 3004 1 x 2 Brick
- x16 98283 1 x 2 Brick with Masonry
- x4 3040 45-Degree Slope Brick
- x4 44728 2 x 2 Bracket
- x3 3622 1 x 3 Brick
- x4 3022 2 x 2 Plate
- x3 3010 1 x 4 Brick
- x3 3021 2 x 3 Plate
- x4 54200 30-Degree Slope Brick
- x1 41539 8 x 8 Plate
- x5 3023 1 x 2 Plate
- x2 98138 1 x 1 Round Tile
- x1 95330pb01 Minifigure Axe

what you'll need

Figure 10-3 shows the LEGO parts used to build the fireplace. None of these parts are in the MINDSTORMS EV3 #31313 set.

The BlinkM will be powered by a 9V battery box using the parts shown in Figure 10-4.

In addition to these LEGO parts, you'll need the following:

- BlinkM programmable LED made by ThingM (*https://thingm.com/*), available at SparkFun Electronics (*https://www.sparkfun.com/*), Digi-Key Electronics (*http://digikey.com/*), and Amazon (*https://www.amazon.com/*)

- LinkM USB programmer for the BlinkM, available at the same stores as the BlinkM

- Zener diode for 5V output, such as the 1N4733A (also used in Chapter 2)

- 47-ohm, 1/2-watt resistor

- 9V battery

- Two jumper wires with female connector ends, preferably one black and one red

- Sugru moldable glue (one 5-gram, single-use pack)

- Soldering iron and solder

- Diagonal cutters

- Wire strippers

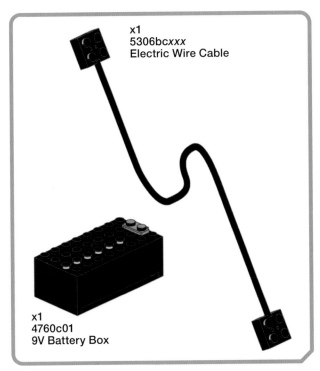

x1
5306bcxxx
Electric Wire Cable

x1
4760c01
9V Battery Box

FIGURE 10-4: LEGO electrical parts used to power the BlinkM

programming the BlinkM

You can program the BlinkM in several ways, but the simplest is to use the LinkM interface and programming utility made by the manufacturer. Figure 10-5 shows the LinkM, which plugs into your computer's USB port like a thumb drive.

FIGURE 10-5: The LinkM allows you to easily program the BlinkM.

Insert the BlinkM to be programmed into the end of the LinkM, as shown in Figure 10-6. Once you've programmed the BlinkM, the program stays in its memory and plays whenever powered on.

FIGURE 10-6: The LinkM plugs into a computer's USB drive.

To program the BlinkM, download the BlinkM software from the ThingM website at *https://thingm.com/products/blinkm/*. Choose **BlinkMSequencer2** for your computer's operating system. From the download, install the program called *BlinkMSequencer*. Running this program will open a window like the one shown in Figure 10-7.

To make a color program, click one of the 48 slots near the top of the screen and then click a color of your choice. The example in Figure 10-7 has a simple program set to turn the LED yellow on every third slot. The program will loop through this pattern at the speed indicated by the Loop Speed drop-down menu. With a 3-second loop speed, the intermittent yellow flicker looks like the muzzle flash from a machine gun.

Of course, your color program is not limited to just one color. You can design colors by using five menu tabs: Swatches, HSV, HSL, RGB, and CMYK. The Swatches tab presents a grid of colors that you can click to select. The other four menu tabs contain sliders that allow fine control of color and brightness settings, which you will see both on the screen and on the BlinkM LED via the LinkM. Once you've selected colors, select how fast the color pattern repeats under the **Loop Speed** control. Choose from 3 seconds, 30 seconds, 100 seconds, 300 seconds, and 600 seconds.

Figure 10-8 shows an example program for the fireplace display, with variations in color meant to look like flames. To create a similar program, you could take

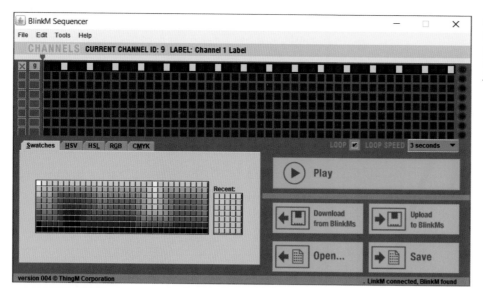

FIGURE 10-7: The BlinkM Sequencer allows you to program a color sequence with 48 steps, shown here with a yellow flash every third step.

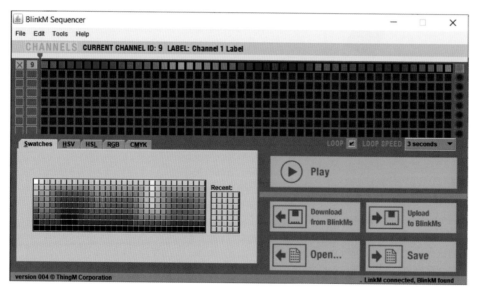

FIGURE 10-8: A color program for the example fireplace cycles through colors to simulate flames.

advantage of a useful tool called Make Gradient, found under the Edit menu. This tool allows you to color two slots. Then it fills in the slots between them with a smooth range of colors. Test your color sequence on the BlinkM by clicking the **Play** button.

When you're done programming, click **Upload to BlinkMs** to set the program in the BlinkM's memory. This program will stay in the BlinkM after you disconnect it from the LinkM. Your BlinkM is now ready to insert into a LEGO invention.

holding the BlinkM

One advantage of the BlinkM is that it's small enough to fit inside LEGO creations. To hold the BlinkM in place with LEGO bricks, you can slide it between the slots of a 1 × 2 tile with grille, as shown in Figure 10-9, which is the base of the fireplace.

Use the following building instructions to build the base. You'll finish the fireplace after powering the BlinkM.

FIGURE 10-9: The base of the LEGO fireplace includes slots to hold the BlinkM.

x2

x2

x1

1

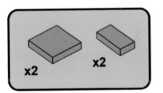

x2

x2

2

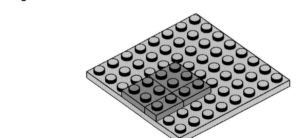

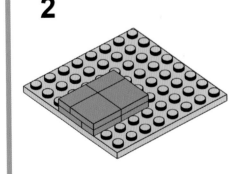

3

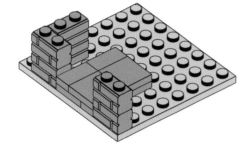

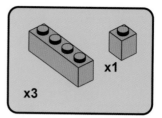

4

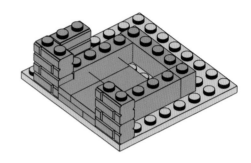

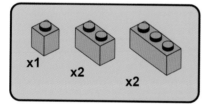

5

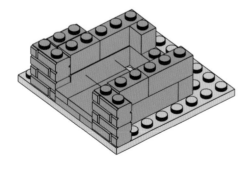

6

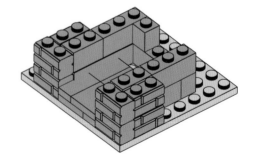

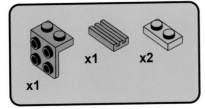

7

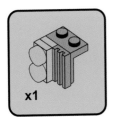

8

9

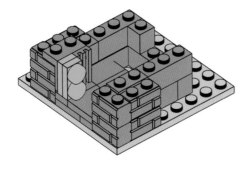

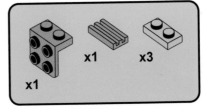

10

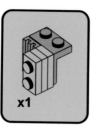

x1

11

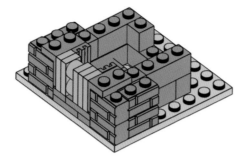

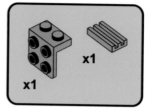

x1

x1

12

x1

13

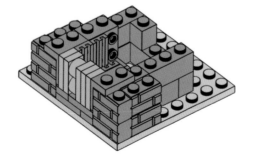

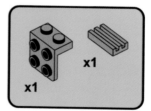

x1

x1

14

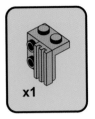

x1

15

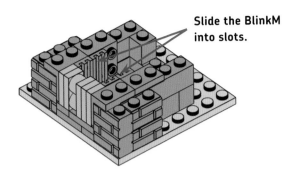

Slide the BlinkM into slots.

powering the BlinkM

The BlinkM connects to power with a pair of pins on its circuit board labeled PWR. One pin is labeled + for the positive connection, and the other is labeled – for the ground connection. *Be sure not to get these two connections confused, since such a voltage reversal could destroy the BlinkM.*

As you did in Chapter 2 for the LEGO-Compatible Laser, you'll have to step the voltage from a 9V battery down to 5.1V by using a resistor and Zener diode circuit. You can read more about stepping down voltage in "connecting the Zener diode and resistor" on page 22. Figure 10-10 shows a diagram of the diode circuit.

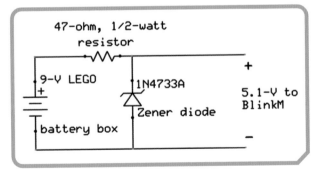

FIGURE 10-10: A Zener diode converts the 9V from a LEGO battery box to the 5.1V needed to power the BlinkM.

To build this circuit, first prepare a LEGO 5306bc electric wire cable, as shown in Figure 10-11, by cutting off the connector on one end (step A), separating the two wires (step B), and stripping the wire ends (step C).

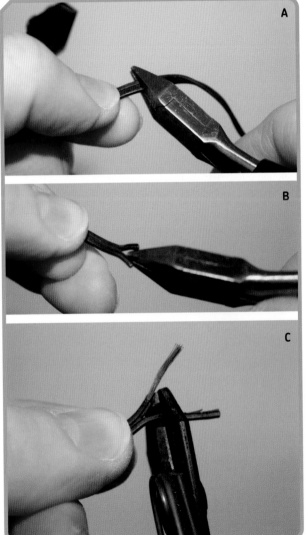

FIGURE 10-11: Prepare a 5306bc electric wire cable by cutting the connector off one end, separating the two wires, and stripping the wire ends.

Then pull the electric wire cable through a 1 × 2 Technic brick with hole, as shown in Figure 10-12. You'll later install this Technic brick as part of the fireplace.

FIGURE 10-12: *Pull the wire through the hole of a 1 × 2 Technic brick.*

Use two jumper wires, preferably one black and one red, both with a female connector on one end. These female connectors will go onto the BlinkM pins, and they're removable, allowing you to take the BlinkM out of the LEGO model in order to reprogram it. Prepare the wires by cutting off a section about 5 cm long. Then strip the insulation from the cut end of the wire using wire strippers, as pictured in Figure 10-13.

FIGURE 10-13: *Strip the insulation off the bare end of the jumper wire.*

Now that you've prepared the jumper wires and electric wire cable, solder together the circuit in Figure 10-10. You'll need to keep track of which wire of the electric wire cable is positive and which is negative. These polarities are shown in Figure 10-14, with the electric wire cable connected to the 9V battery box.

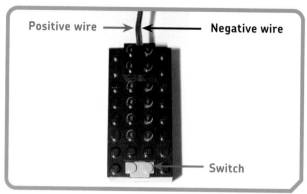

FIGURE 10-14: *A 9V battery box (4760c01) with an electric wire cable (5306bc) attached in this orientation has the positive connection on the left side and the negative connection on the right side.*

Now make the three solder connections shown in Figure 10-15:

Solder joint 1 Three leads consisting of the negative lead from the LEGO battery box, the black jumper wire, and the anode end of the Zener diode. The anode end of the Zener diode is the one without the black band.

Solder joint 2 Three leads consisting of one end of the resistor, the red jumper wire, and the cathode end of the Zener diode. The cathode end of the Zener diode has a black band around it.

Solder joint 3 Two leads consisting of the end of the positive lead from the LEGO battery box and the other end of the resistor.

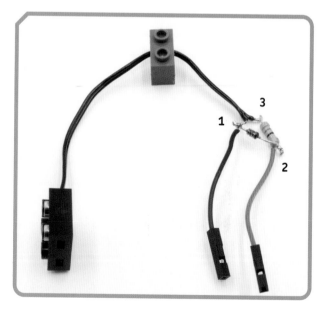

FIGURE 10-15: *Three solder joints connect the BlinkM to its power source.*

Place Sugru moldable glue around the circuit components. This Sugru encapsulation keeps the soldered connections from coming loose and insulates bare wires from each other. Keep the Sugru blob as small as possible so you can easily hide it in your LEGO model.

Before the Sugru hardens, install the power connection by attaching the red jumper wire connector onto the BlinkM's positive (+) terminal and the black jumper wire connector onto the negative (–) terminal. As shown in Figure 10-16, slide the BlinkM into the grille slots, place the 1 × 2 Technic brick into the fireplace base, and tuck the Sugru-encapsulated circuit into the cavity behind the BlinkM.

Finally, use leftover Sugru around the hole in the 1 × 2 Technic brick to keep the wire cable from wriggling around. Let the Sugru cure overnight and then power on your BlinkM.

finishing the fireplace

After making the power connections, finish building the fireplace with these building instructions. The building steps begin by showing the placement of the 1 × 2 Technic brick with hole, through which you fed the wire cable.

FIGURE 10-16: Insert the power connections into the fireplace's base.

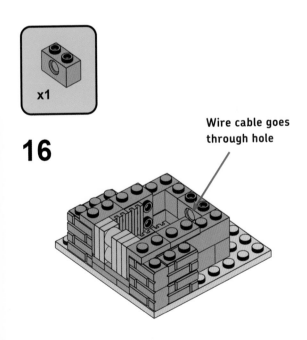

x1

Wire cable goes through hole

16

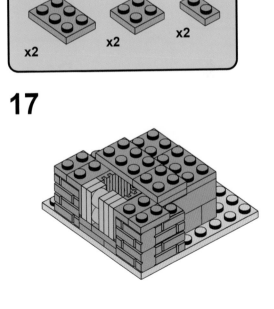

x2   x2   x2

17

x4

18

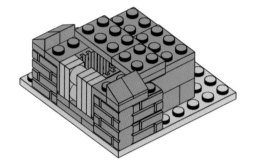

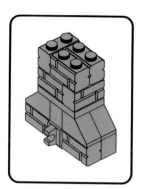

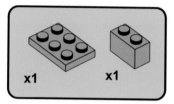

x1   x1

19

x4

20

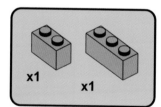

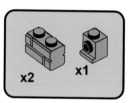

x1   x1

21

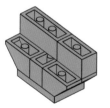

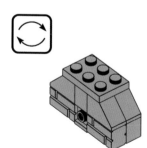

x2   x1

22

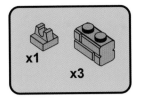

x1 x3

23

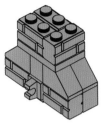

x3

24

Add height to the chimney, if desired, by repeating the pattern.

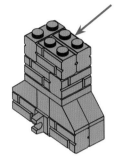

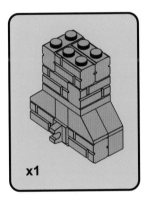

x1

25

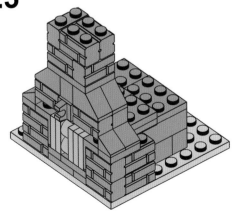

x1

26

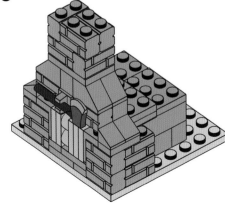

ideas for going further

You can power up to three BlinkMs from a single 9V battery box by stacking the electric wire cable connectors on top of each other. You would have to make a Zener diode circuit for each BlinkM. The power draw increases with the addition of every BlinkM. Three BlinkMs should be able to run off a single alkaline battery for 2.5 hours and off a lithium battery for 6 hours.

Improve your display by adding a mist generator to BlinkM lighting. The mist will take on the colors created by the BlinkM, creating a glowing fog. A mist generator is available as a Grove device from Seeed Studio (*https://www.seeedstudio.com/*). It is compatible with the Grove Sensor Adapter used in Chapter 5, making it possible to control with an EV3 Intelligent Brick. You can see an example of the mist generator in use at *https://hightechlego.com/*.

what you learned

In this chapter, you learned how to create an LED display that changes color by using a BlinkM. You know how to program the LED colors to make a custom pattern and set the speed at which the color pattern will play. After you program the device on your computer, the BlinkM will play the color program as long as power is applied. You also built a power supply for the BlinkM by using a LEGO 9V battery box and then embedded it in a fireplace LEGO model.

11

the laser light show

In Chapter 3, you saw one possible use of a laser: building an alarm system by taking advantage of the way a laser's output generates a narrow beam. In this chapter, you'll use the laser beam to draw geometric patterns with spinning mirrors. By spinning the mirrors at different speeds and in different directions, you can create a variety of patterns. This project reuses the LEGO-Compatible Laser built in Chapter 2.

Keep in mind that lasers pose a safety hazard. Before you begin this project, read "Safety Warnings" on page 17. When you bounce a laser beam off a mirror, the beam can go in unexpected directions, so take care when turning on this device. If you're using the laser you made in Chapter 2, accidentally shining the laser beam briefly into your eyes shouldn't damage them, but it's a good safety practice to always point a laser beam away from people, including yourself.

what you'll need

Figures 11-1 and 11-2 show the LEGO parts used to build the Laser Light Show. Some of these parts, marked with an asterisk, are not in the MINDSTORMS EV3 #31313 set. Note that you'll need two EV3 Medium Motors for this project, whereas only one comes with the MINDSTORMS EV3 #31313 set.

In addition to the LEGO parts in those two figures, you'll need the following:

* The LEGO-Compatible Laser built in Chapter 2, including its battery box power source.

* Two square front-surface metallic mirrors with widths of 25.4 mm (1 inch). These mirrors are available at Thorlabs (*https://www.thorlabs.com/*, part number ME1S-G01), Edmund Optics (*https://www.edmundoptics .com/*, part number 69-246), Newport (*https://www .newport.com/*, part number 10SJ00ER.3), or Esco Optics (*https://escooptics.com/*, part number D610010).

* Cyanoacrylate glue, such as Super Glue, Krazy Glue, or Gorilla Glue (a few drops from a small tube).

* Latex or nitrile gloves (not powder coated).

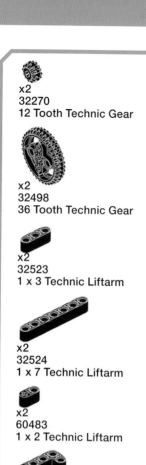
x2
32270
12 Tooth Technic Gear

x2
32498
36 Tooth Technic Gear

x2
32523
1 x 3 Technic Liftarm

x2
32524
1 x 7 Technic Liftarm

x2
60483
1 x 2 Technic Liftarm

x2
32140
Technic 2 x 4 L-Shape Liftarm

x6
3001
2 x 4 Brick*

x6
3020
2 x 4 Plate*

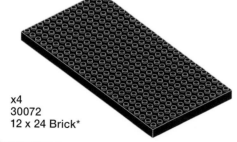
x4
30072
12 x 24 Brick*

x4
3795
2 x 6 Plate*

x4
3701
1 x 4 Technic Brick with Holes*

x2
3702
1 x 8 Technic Brick with Holes*

x12
43093
Technic Axle Pin

x2
32184
Technic Axle and Pin Connector

x2
42003
Technic Axle and Pin Connector

x22
2780
Technic Pin with Friction Ridges

x1
64178
5 x 11 Open Center Frame Liftarm

x4
48989
Technic 4-Pin Connector

x2
99008
Technic Axle with Center Stop

x2
55013
Technic Axle 8 with Stop

x2
60485
Technic Axle 9

x4
3713
Technic Bush

x6
4265c
Technic Bush 1/2 Smooth

x2
3649
40 Tooth Technic Gear*

x4
4274
Technic Pin 1/2*

x1
87081
Locking Turntable*

FIGURE 11-1: LEGO bricks used in the Laser Light Show

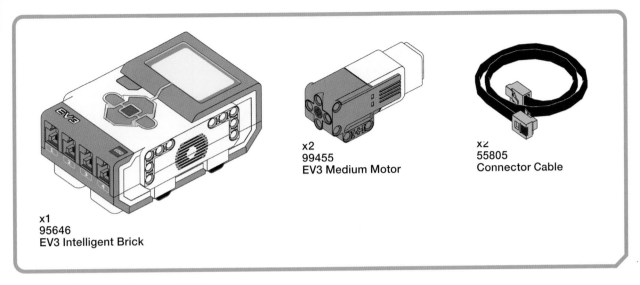

FIGURE 11-2: EV3 parts used in the Laser Light Show

x2
99455
EV3 Medium Motor

x2
55805
Connector Cable

x1
95646
EV3 Intelligent Brick

front-surface mirrors

The front-surface mirrors needed for optics and laser experiments aren't the same as the rear-surface mirrors you probably have in your home. While manufacturers make all mirrors by coating a piece of glass with a thin layer of metal, a household mirror is made by coating the rear surface of the glass and then painting over the metal with an opaque layer. So, when looking at yourself in a bathroom mirror, you're looking at your reflection through a thick layer of glass. This design protects the metal layer that does the reflecting. Otherwise, the mirror would be easily scratched and tarnished in daily use.

The problem with using a rear-surface mirror to reflect a laser beam is that both the glass and the metal surface behind it will reflect the beam. Not only that, but the two surfaces will bounce light between them, producing a jumbled mess to work with (Figure 11-3).

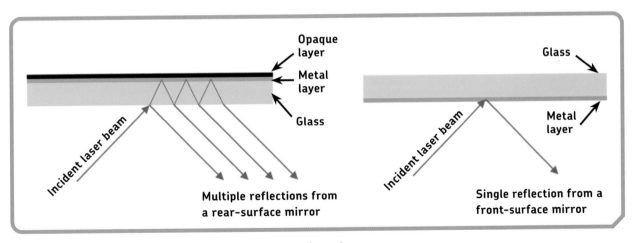

Opaque layer
Metal layer
Glass
Incident laser beam
Multiple reflections from a rear-surface mirror

Glass
Metal layer
Incident laser beam
Single reflection from a front-surface mirror

FIGURE 11-3: Two types of mirrors: rear-surface (at left) and front-surface (at right)

Figure 11-4 shows the reflection of the LEGO-Compatible Laser of Chapter 2 from a rear-surface mirror, with the many reflected spots that result. By contrast, a front-surface mirror has reflective metal on the front of the glass. Front-surface mirrors are much easier to work with, as they create only one spot.

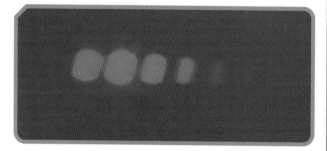

FIGURE 11-4: *A laser beam reflected from a rear-surface mirror will have multiple spots.*

Be careful when handling your front-surface mirror since the metal surface is easily smudged or scratched. Most of these mirrors come in a foam-lined box and are wrapped in a soft cloth that won't scratch the mirror. Save the box and cloth for storing your mirrors when they're not in use. It's best to use latex or nitrile gloves when working with the mirror to avoid smudging its surface with finger oils. Also, hold the mirror by its rear surface or edges.

the law of reflection

When mirrors reflect laser beams, they follow the *law of reflection*, which states that the angle of reflection is equal to the angle of incidence. The *angle of reflection* is the angle at which the bounced laser beam leaves the mirror. The *angle of incidence* is the angle at which the incoming laser beam hits the mirror. Figure 11-5 represents these angles as Θ_i and Θ_r, respectively. The imaginary line that separates the angles, called a *normal*, is perpendicular to the mirror's surface. By changing a laser beam's angle of incidence, we change its angle of reflection as well, deflecting the laser beam (Figure 11-6).

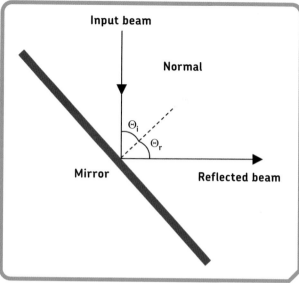

FIGURE 11-5: *The law of reflection shows that the angle of incidence is equal to the angle of reflection.*

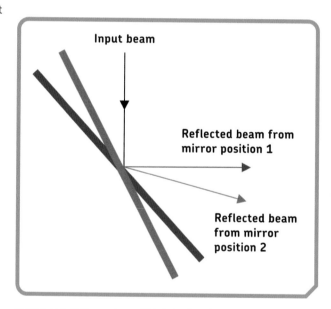

FIGURE 11-6: *Tilting a mirror will deflect a laser beam.*

You can vary the angle of incidence by tilting the mirror a little and then spinning it on a motor. If you project the resulting pattern onto a wall, it should look like an oval. A single oval isn't very interesting to look at, but you can build off of it to create many interesting shapes by bouncing the beam off a second spinning mirror, as you'll do in this chapter.

building the laser light show

The Laser Light Show, shown in Figure 11-7, consists of three subsystems: the baseplate, the mirror spinners, and the EV3 Intelligent Brick mount.

FIGURE 11-7: *The Laser Light Show uses spinning mirrors to draw geometric patterns with laser light.*

building the LEGO mirror mount

To hold each mirror steady and mount it to LEGO pieces, you'll glue it to a 40-tooth Technic gear. This gear has a useful center axle connection and pin connections that tilt the mirror to account for the law of reflection (discussed in "The Law of Reflection" on page 138). The angle of tilt you need is small—in the 1- to 4-degree range—and the design shown here uses a tilt of about 2.5 degrees.

The pin connections need to be secure, since the mirrors will be spinning fast. Place a couple of drops of glue on two Technic pins 1/2 before inserting them into the holes of the gear, as shown in Figure 11-8.

FIGURE 11-8: *Glue Technic pins 1/2 into a pair of outermost holes of a 40-tooth Technic gear.*

Let the glue dry and then turn the Technic gear over. Notice that the pins poke through the gear at a height of about a millimeter. Glue the mirror onto this side (the side on which the pins barely protrude) of the gear, allowing one side of the mirror to rest against the protruding Technic pins, as shown in Figure 11-9. The other end of the mirror should sit against the gear's surface. By attaching the mirror at these two edges, you've created the small tilt you need.

FIGURE 11-9: *Glue the mirror onto the side of the gear that has the Technic pins barely protruding.*

Now place glue around all four edges of the mirror, taking care to avoid getting glue on the mirror's surface. Repeat the gluing procedure for a second mirror and 40-tooth Technic gear.

building the baseplate

The Laser Light Show mirrors will spin rapidly, producing torque that requires a sturdy base. Figure 11-10 shows the baseplate design, which includes places to attach the mirror spinners and EV3 Intelligent Brick mount that you'll build in a later section. You'll also attach the LEGO-Compatible Laser to the baseplate.

Follow the steps shown here to build the baseplate.

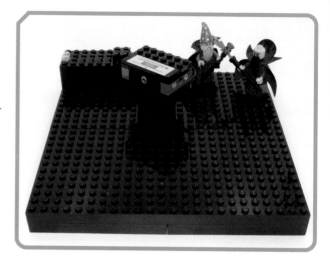

FIGURE 11-10: The baseplate is built from 12 × 24 bricks.

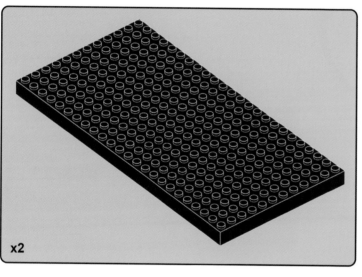

x2

1

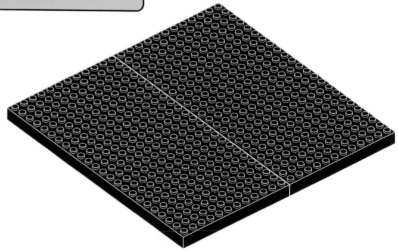

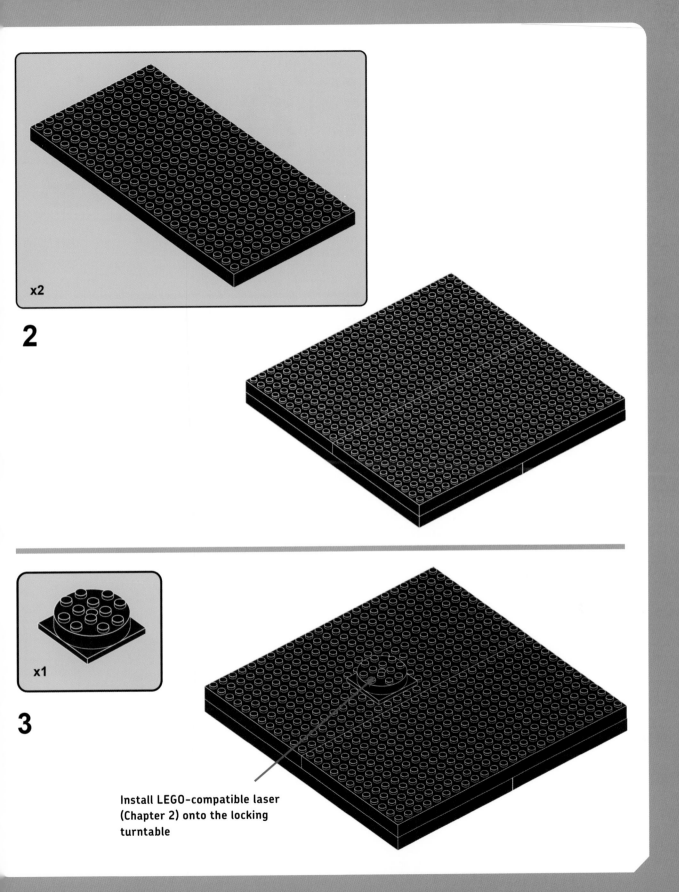

2

x2

3

x1

Install LEGO-compatible laser
(Chapter 2) onto the locking
turntable

building the mirror spinners

The mirror spinners use EV3 Medium Motors (Figure 11-11). You'll build two of them and then attach the gear-mounted mirrors you glued together earlier. A gear arrangement speeds up the rotation of the mirror: a large Technic 36-tooth gear connects to the EV3 Medium Motor, which, in turn, spins a Technic 12-tooth Gear, increasing the rotation speed by a factor of three.

Follow these instructions to build each mirror spinner.

FIGURE 11-11: A mirror spinner, with the mirror removed to reveal the spinning mechanism

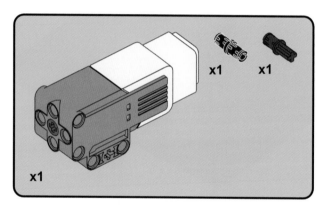

1

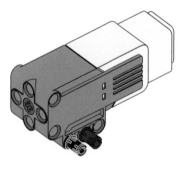

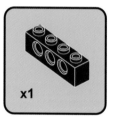

2

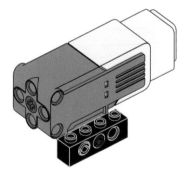

x1   x1

3

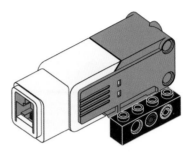

4

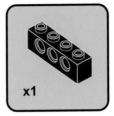

x1

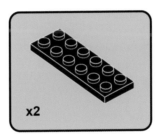

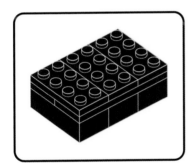

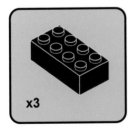

x3

5

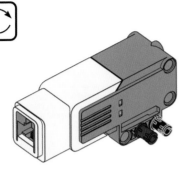

x2

6

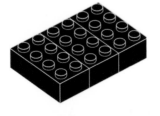

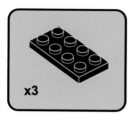

x3

x1

7

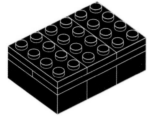

8

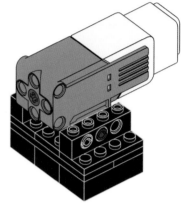

x2   x2

9

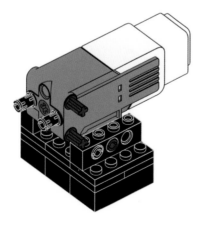

x1   x1

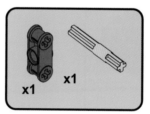

10

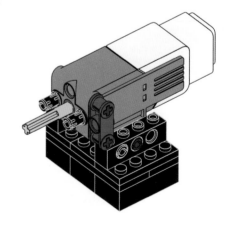

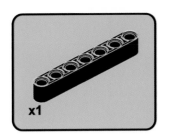

11

12

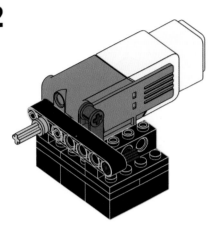

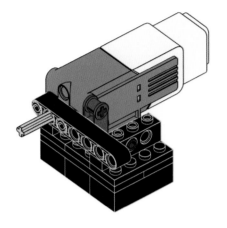

13

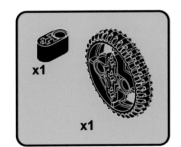

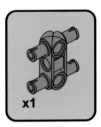

14

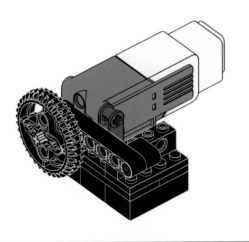

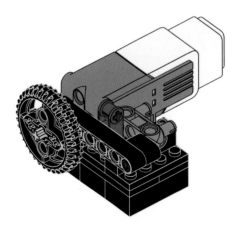

15

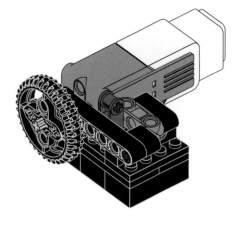

16

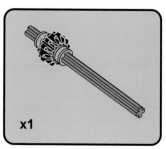

17

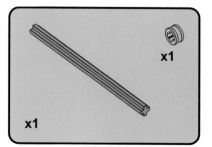

18

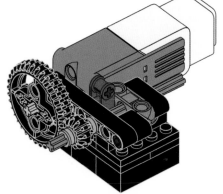

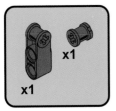

19

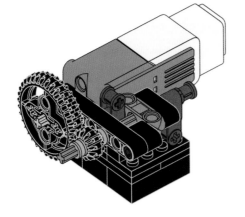

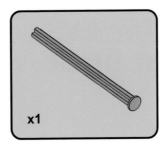

20

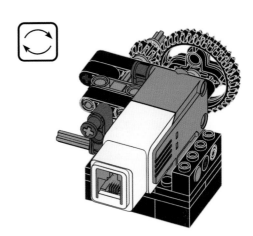

21

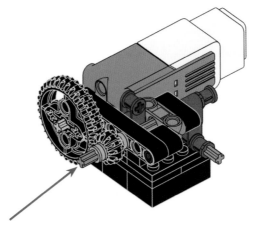

Install the gear-mounted
mirror onto the axle.

building the EV3 Intelligent Brick mount

The EV3 Intelligent Brick controls the motors with a program you'll write later. In this section, you'll build a mount for the EV3 Intelligent Brick, shown in Figure 11-12, so that it can fit onto the baseplate. Use the following building instructions to create it.

FIGURE 11-12: *Add bricks to the EV3 Intelligent Brick to attach it to the baseplate.*

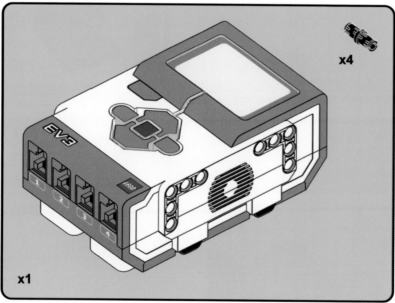

x4

x1

1

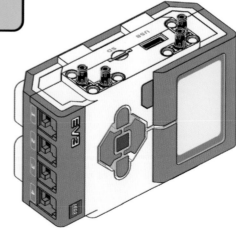

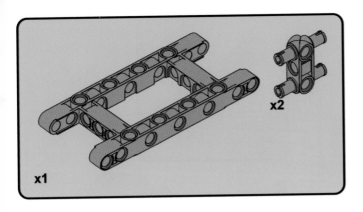

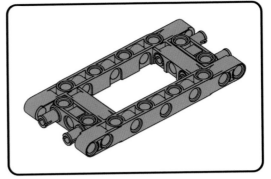

2

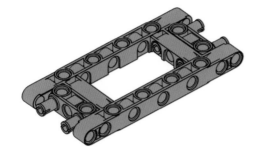

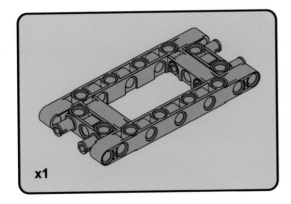

3

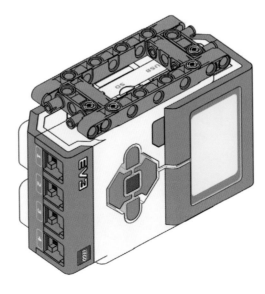

x5

4

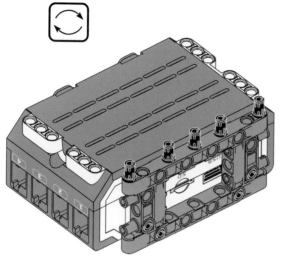

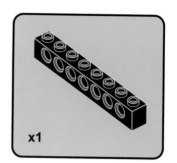

x1

5

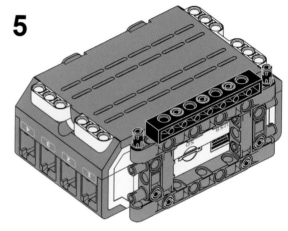

x3

6

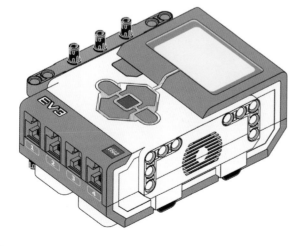

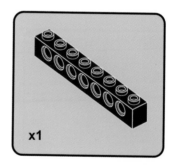

x1

7

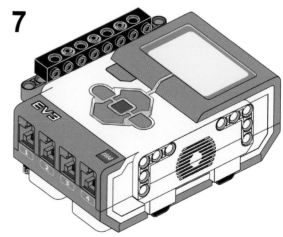

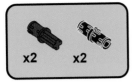

8

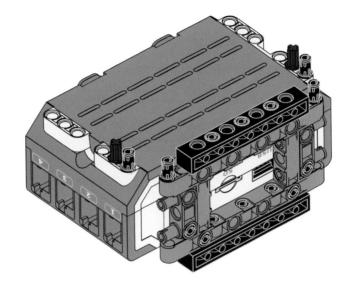

9

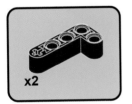

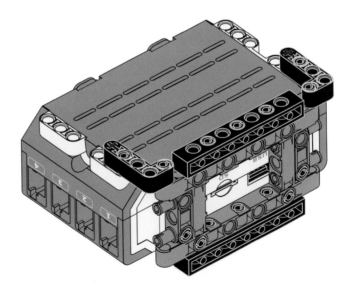

assembling the laser light show

Now that you've built all the parts for the Laser Light Show, assemble them on the baseplate. Figure 11-13 provides a map of where to install the LEGO-Compatible Laser (along with its battery box power supply), mirror spinners, and EV3 Intelligent Brick. Once you've installed all the subassemblies, turn on the LEGO-Compatible Laser and rotate it on its turntable so it points to the closer mirror. The laser beam should hit this first mirror at an angle, and the reflection from the first mirror should hit the second mirror near its

center (Figure 11-14). Finally, the reflection from the second mirror should pass through the gap between the LEGO-Compatible Laser and EV3 Intelligent Brick.

To connect the EV3 Intelligent Brick to the two EV3 Medium Motors, place a connector cable between port A on the EV3 Intelligent Brick and the EV3 Medium Motor closer to the laser. Insert another connector cable between port D on the EV3 Intelligent Brick and the EV3 Medium Motor that spins the second mirror.

FIGURE 11-13: Attach the LEGO-Compatible Laser (along with its battery box power supply), mirror spinners, and EV3 Intelligent Brick to the baseplate.

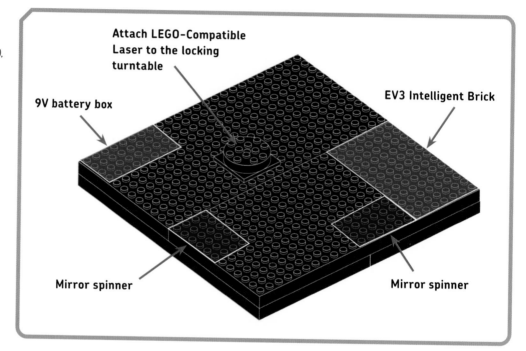

Attach LEGO-Compatible Laser to the locking turntable

9V battery box

EV3 Intelligent Brick

Mirror spinner

Mirror spinner

FIGURE 11-14: Aim the LEGO-Compatible Laser so it hits the first mirror at its right edge.

writing the
software

The program shown in Figure 11-15 spins the EV3 Medium Motors, varying the motor directions and speeds to change the light pattern created.

Use the following blocks to write the program.

1. **Loop block**   This block continuously repeats the motor instructions.

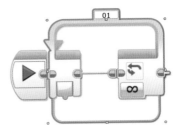

Place a Loop block after the program Start.

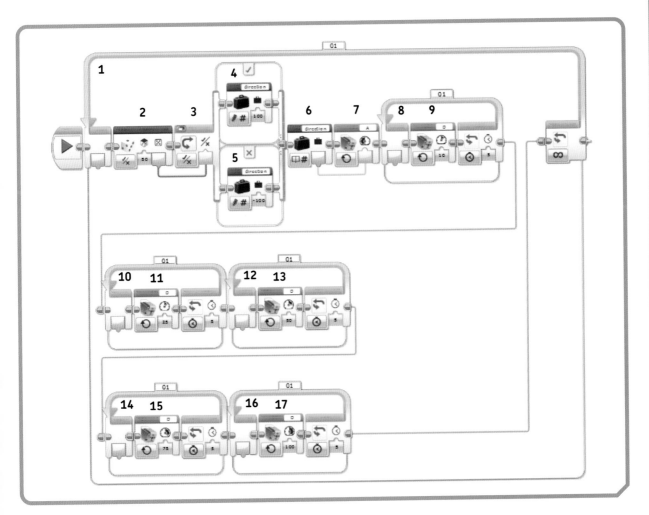

FIGURE 11-15: *The program varies the motor directions and speeds.*

2. **Random block** This block chooses either true or false. This random decision will determine the direction of one of the motors.

Insert a Random block inside the Loop and set **Random** to **Logic** and **Probability of True** to **50**.

3. **Switch block** This block creates two possible values for a variable that will be assigned based on the result of the Random block.

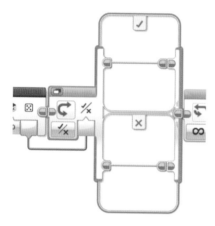

Place a Switch block after the Random block and set **Switch** to **Logic**. Connect a wire from the Random block's **Value** output to the Switch's **Logic** input.

4. **Variable block** This assigns a value of **100** to a variable called direction. We'll use this value later as an input to a motor.

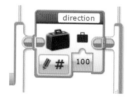

Place a Variable block inside the Switch block's true path. Set **Variable** to **Write-Numeric** and enter **100** into **Value**. Name the variable **direction**.

5. **Variable block** This assigns a value of **–100** to the direction variable. This value is the negative of the value assigned in block 4. Therefore, the true and false paths give opposite values to the variable.

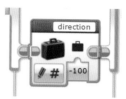

Place a Variable block inside the Switch block's false path. Set **Variable** to **Write-Numeric** and enter **–100** into **Value**. Name the variable **direction**.

6. **Variable block** This block reads the value stored in the direction variable.

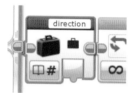

Insert a Variable block after the Switch block and set **Variable** to **Read-Numeric**. Name the variable **direction**.

7. **Medium Motor block** This block turns on the Medium Motor connected to port A. This motor controls the position of the first mirror that the laser beam hits. Depending on the value of the direction variable, the motor's power setting will be either –100 or 100, resulting in the highest speed setting of the motor, with a spin of either clockwise or counterclockwise.

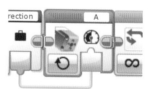

Connect a Medium Motor block following the Variable block. Set **Medium Motor** to **On** and **Port** to **A**. Connect a wire from the **Power** input to the Variable block's **Value** output.

8. **Loop block**   This block sets an action that repeats for a duration of 5 seconds.

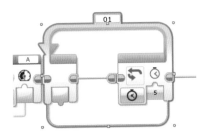

Place a Loop block after the Medium Motor block. Set **Loop** to **Time** and **Seconds** to **5**.

9. **Medium Motor block**   This block activates the motor connected to port D, which controls the second mirror that the laser beam hits. Because this motor activation occurs inside a timed loop, the motor will stay on for the loop's duration. The power of the motor is set to 10, a slow rotation speed.

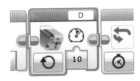

Place a Medium Motor block inside the Loop block. Set **Medium Motor** to **On**, **Port** to **D**, and **Power** to **10**.

10–11. **Motor in Loop**   This repeats steps 8 and 9, except it sets the motor speed to 25.

Duplicate the blocks of steps 8 and 9 and place them after block 8. Change the Medium Motor block's **Power** to **25**.

12–13. **Motor in Loop**   This repeats steps 8 and 9, except it sets the motor speed to 50.

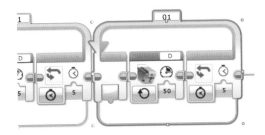

Duplicate the blocks in steps 8 and 9 and place them after block 10. Change the Medium Motor block's **Power** to **50**.

14–15. **Motor in Loop**   This repeats steps 8 and 9, setting the motor speed to 75.

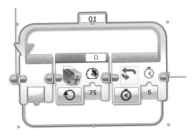

Duplicate the blocks of steps 8 and 9 and place them after block 12. Change the Medium Motor block's **Power** to **75**.

16–17. **Motor in Loop**   This repeats steps 8 and 9 one last time, setting the motor speed to 100.

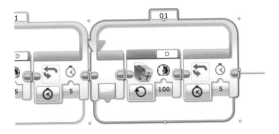

Duplicate the blocks in steps 8 and 9 and place them after block 14. Change the Medium Motor block's **Power** to **100**.

using the laser light show

Turn on the LEGO-Compatible Laser and point the laser beam at a wall. When you activate the program from Figure 11-15, the Laser Light Show should project geometric patterns. Every 5 seconds, the speed of the motor should change, resulting in a variety of patterns, such as spinning circles, rotating lines, and rosettes. Watch videos of the light show in action at *http://hightechlego.com/*.

Figure 11-16 shows the laser pattern at a distance of about 25 cm from the mirrors. The pattern will get much bigger if projected over a longer distance.

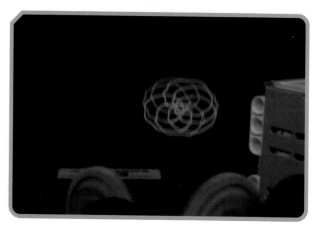

FIGURE 11-16: *The laser beam pattern is projected through the gap between the LEGO-Compatible Laser and the EV3 Intelligent Brick.*

ideas for going further

Alter the program to experiment with new patterns, changing the motor speeds and directions. Currently, the software changes only the first motor's direction and the second motor's speed. Try modifying the software to change the second motor's direction instead, or the first motor's speed. Even better, change the speed and direction of both mirrors.

Another technique you could use to change the laser pattern is to offset a mirror by a different angle. To do this, remove one of the mirrors from its axle, rotate it 90 degrees, and put it back on the axle. Doing this can create dramatic differences in the resulting patterns.

You can also create different geometric patterns by changing the tilt angle of the second mirror (though this would require buying and installing another mirror). For example, you could glue the mirror to the side of the Technic gear where the Technic pins 1/2 stick out farther, rather than the side where the pins barely protrude through the gear, giving it a steeper tilt. You'll need to move the EV3 Intelligent Brick out of the beam path for this to work. Alternatively, experiment with gear sizes to give ratios other than the 3:1 ratio used in this chapter, which would result in different spin speeds. A higher ratio generates a faster speed.

Laser patterns are especially mesmerizing when viewed through fog, so take the Laser Light Show outside if there's a foggy night in your weather forecast. You can also use fog machines or spray cans to enhance the light show.

The EV3 Intelligent Brick automates the motor control used here, but you could add an interactive control that lets people change the laser patterns by manipulating the Intelligent Brick's front panel buttons, EV3 Touch Sensor, or EV3 Ultrasonic Sensor. The EV3 Ultrasonic Sensor, which could let you change the pattern by moving your hand varying distances from the sensor, would be particularly interesting.

what you learned

In this chapter, you built a Laser Light Show that projects geometric patterns using the LEGO-Compatible Laser created in Chapter 2. You learned how to use mirrors to direct a laser beam, taking advantage of the law of reflection and working with specialized front-surface mirrors. The laser show's geometric patterns result from the spinning of two mirrors. Altering mirror angles, spin speeds, and spin directions can create different geometric patterns.

12

the infrared thermometer and cannon

People often interpret the term *radiation* as referring solely to the harmful radioactive emissions from nuclear reactors or atomic bombs. But all objects emit radiation in the form of heat. Radiation, in general scientific terms, refers to any energy coming from an object. For example, you can feel infrared radiation from the sun when you're outdoors.

Sensors called *radiometers* can measure infrared radiation and even quantify the surface temperature of an object. Satellites use radiometers to measure temperatures related to the earth's clouds, layers of atmosphere, and oceans. The military also uses this sensing technology for heat-seeking missiles that home in on the hot engines of aircraft. In this chapter, you'll add a radiometer to a couple of LEGO inventions. The first project, useful as a scientific instrument, is a handheld Remote Sensing Thermometer that you can point at an object to measure its surface temperature without touching the object. The second project is a tabletop Heat-Seeking Cannon that shoots LEGO studs at hot objects within a certain range.

working with infrared radiation

The radiometer used for the projects in this chapter is the IR Temperature Sensor from Mindsensors (Figure 12-1). The circular part at the top of the device collects infrared radiation.

FIGURE 12-1: *The IR Temperature Sensor made by Mindsensors detects the infrared radiation emitted by objects.*

The thermal energy emitted by warm objects is mostly in the far-infrared part of the electromagnetic spectrum, a region characterized by much longer wavelengths than the near-infrared light described in Chapter 5. Therefore, the sensing technology used in the IR Temperature Sensor is different from the technology for detecting near-infrared light. You can't see far-infrared radiation with a night vision monocular device or smartphone camera, as you did in Chapter 5.

what you'll need

Figures 12-2 and 12-3 show the LEGO components used to build both projects. All LEGO parts are in the MINDSTORMS EV3 #31313 set, except for the projectile launchers and 1×1 round plates marked by an asterisk. The 1×1 round plates are the stud ammunition shot from the Heat-Seeking Cannon.

In addition to these LEGO parts, you'll need the IR Temperature Sensor from Mindsensors (*https://www .mindsensors.com/*).

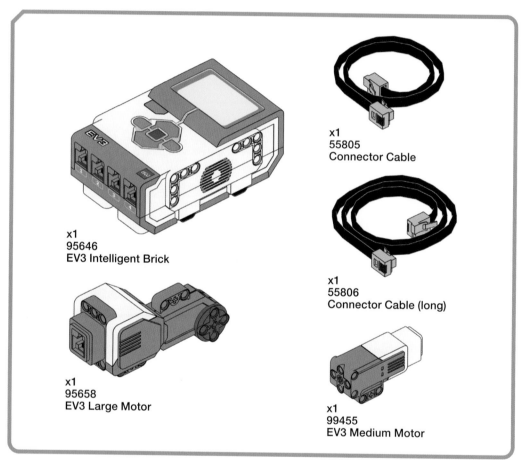

x1
95646
EV3 Intelligent Brick

x1
55805
Connector Cable

x1
55806
Connector Cable (long)

x1
95658
EV3 Large Motor

x1
99455
EV3 Medium Motor

FIGURE 12-2: EV3 parts used for projects with infrared radiation

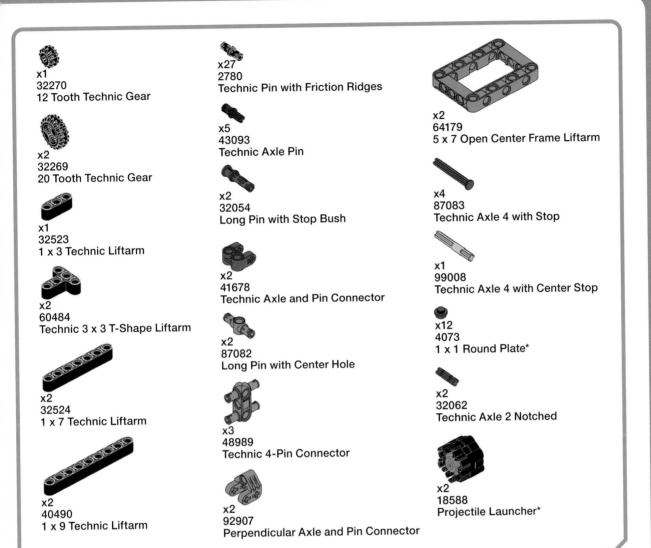

x1
32270
12 Tooth Technic Gear

x2
32269
20 Tooth Technic Gear

x1
32523
1 x 3 Technic Liftarm

x2
60484
Technic 3 x 3 T-Shape Liftarm

x2
32524
1 x 7 Technic Liftarm

x2
40490
1 x 9 Technic Liftarm

x27
2780
Technic Pin with Friction Ridges

x5
43093
Technic Axle Pin

x2
32054
Long Pin with Stop Bush

x2
41678
Technic Axle and Pin Connector

x2
87082
Long Pin with Center Hole

x3
48989
Technic 4-Pin Connector

x2
92907
Perpendicular Axle and Pin Connector

x2
64179
5 x 7 Open Center Frame Liftarm

x4
87083
Technic Axle 4 with Stop

x1
99008
Technic Axle 4 with Center Stop

x12
4073
1 x 1 Round Plate*

x2
32062
Technic Axle 2 Notched

x2
18588
Projectile Launcher*

FIGURE 12-3: Other LEGO parts used in this chapter

the remote sensing thermometer

The Remote Sensing Thermometer allows you to measure the surface temperature of an object by pointing it at the object from a distance of about 1 meter. This keeps you from having to physically touch the object you want to measure, which is convenient if you can't reach the object or if the object may be too hot to safely touch.

Figure 12-4 shows the Remote Sensing Thermometer, built by attaching the Mindsensors IR Temperature Sensor to the EV3 Intelligent Brick. When you point the Remote Sensing Thermometer at an object, the EV3 Intelligent Brick's front panel will display the object's surface temperature, as well as additional information about the ambient room temperature, measured by a conventional thermometer embedded in the IR Temperature Sensor. With this additional information, you can compare the temperature of the object to the surrounding air temperature.

FIGURE 12-4: *The Remote Sensing Thermometer measures the surface temperature of whatever object you point it at.*

building the remote sensing thermometer

Use the following building instructions to create the Remote Sensing Thermometer.

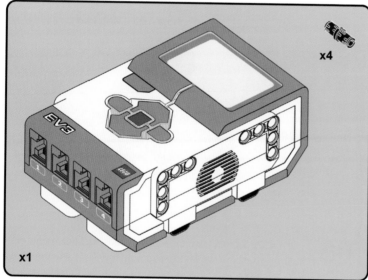

x4

x1

1

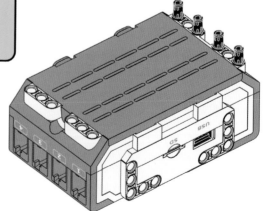

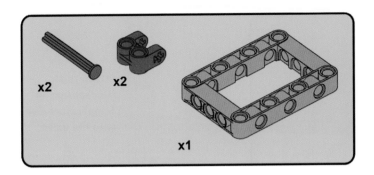

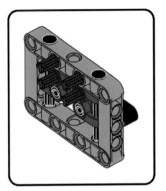

2

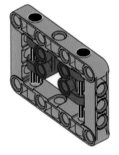

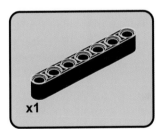

3

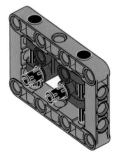

4

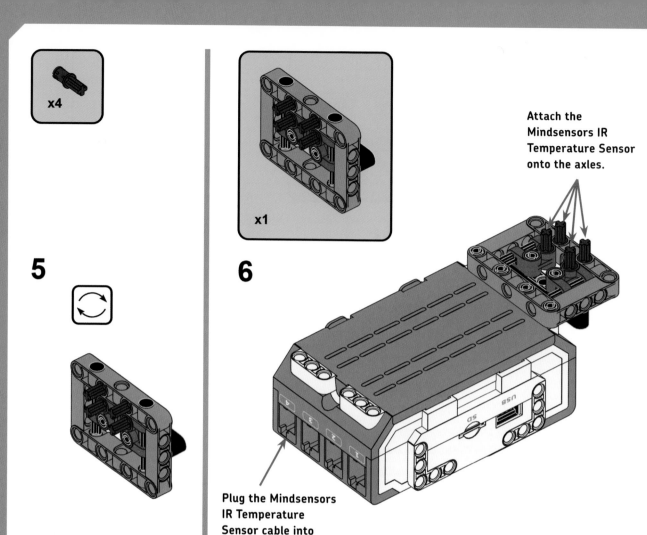

x4

x1

5

6

Attach the
Mindsensors IR
Temperature Sensor
onto the axles.

Plug the Mindsensors
IR Temperature
Sensor cable into
port 4.

writing the software

The EV3 program will read temperature measurements from the IR Temperature Sensor and display them on the front panel of the EV3 Intelligent Brick. Before you begin, download the IR Temperature Sensor block from

Mindsensors at *http://www.mindsensors.com/products /170-ir-temperature-sensor-for-ev3-or-nxt*. Import the block into the MINDSTORMS EV3 programming environment by using the instructions in "writing the software for the covert Morse code transmitter" on page 61. Figure 12-5 shows the EV3 program for the Remote Sensing Thermometer.

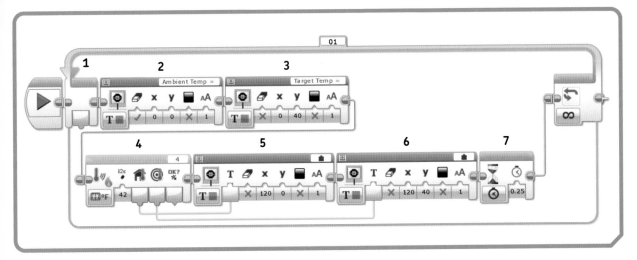

FIGURE 12-5: The EV3 program for the Remote Sensing Thermometer

The programming blocks perform the following functions:

1. **Loop block** This block repeatedly measures the temperature.

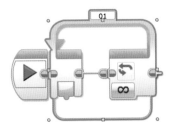

Place a Loop block after the program Start.

2. **Display block** This block prints a label on the front panel of the EV3 Intelligent Brick to describe a measurement that the sensor will make.

Insert a Display block inside the Loop block. Set **Display** to **Text—Pixels**, **Text** to **Ambient Temp =**, **Clear Screen** to **True**, **x** to **0**, **y** to **0**, **Color** to **False**, and **Font** to **1**.

3. **Display block** This prints another label on the front panel for a measurement that the sensor will make.

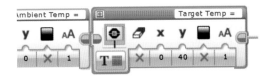

Place a new Display block after the existing Display block. Set the new block's parameters for **Display** to **Text—Pixels**, **Text** to **Target Temp =**, **Clear Screen** to **False**, **x** to **0**, **y** to **40**, **Color** to **False**, and **Font** to **1**.

4. **Mindsensors IR Thermometer block** This takes measurements from the IR Temperature Sensor. This block, downloaded from mindsensors.com, should appear in the Sensor tab after you've installed it into the MINDSTORMS EV3 programming environment.

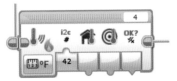

Place a Mindsensors IR Thermometer block after the Display block. Set **Mindsensors IR Thermometer** to **Measure-Fahrenheit** and **Port** to **4**. If you prefer to use the Celsius temperature scale, choose the menu option for measuring in Celsius.

5. **Display block**   This prints the ambient temperature measured by the IR Temperature Sensor on the front panel of the EV3 Intelligent Brick. *Ambient temperature* is the measure of the general environment that you're in. If you're indoors, we call this the *room temperature*. The measurement result will appear on the screen after the label programmed in block 2.

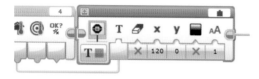

Place a Display block after the Mindsensors IR Thermometer block. Set **Display** to **Text—Pixels**, **Text** to **Wired**, **Clear Screen** to **False**, **x** to **120**, **y** to **0**, **Color** to **False**, and **Font** to **1**. Run a wire from **Text** input to the **Ambient Fahrenheit Temperature** output of the Mindsensors IR Thermometer block.

6. **Display block**   This prints the target temperature measured by the IR Temperature Sensor on the front panel of the EV3 Intelligent Brick. The *target temperature* is the surface temperature of the specific object at which you're pointing the Remote Sensing Thermometer. The measurement results will be printed on the screen after the label programmed in block 3.

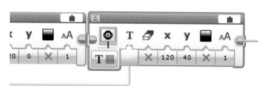

Place a new Display block after the previous Display block. Set **Display** to **Text—Pixels**, **Text** to **Wired**, **Clear Screen** to **False**, **x** to **120**, **y** to **40**, **Color** to **False**, and **Font** to **1**. Run a wire from the **Text** input to the **Target Fahrenheit Temperature** output on the Mindsensors IR Thermometer block.

7. **Wait block**   This pauses action for a short period to prevent data from becoming confused as information is passed between the IR Temperature Sensor and the EV3 Intelligent Brick.

Place a Wait block after the Display block. Set **Wait** to **Time** and **Seconds** to **0.25**.

using the remote sensing thermometer

Once you've programmed the EV3 Intelligent Brick, it's time to use your new temperature measurement tool. Two temperatures should appear on the front panel of the EV3 Intelligent Brick: the ambient room temperature and the target temperature of the object at which you point the Remote Sensing Thermometer. Notice that if you point the sensor at a wall, the ambient temperature and target temperature should be about the same. Next, try pointing the Remote Sensing Thermometer at your hand from about 10 cm away. You should see the target temperature rise several degrees above ambient room temperature. Note that the Remote Sensing Thermometer measures the surface temperature of your hand, not the much warmer internal temperature of your body.

The most interesting application for the Remote Sensing Thermometer is measuring objects that you can't or shouldn't touch, like the fire shown in Figure 12-6. On the cold winter day that this picture was taken, the ambient temperature measured 36 degrees Fahrenheit (2.2 degrees C), while the fire's target temperature measured upward of 350 degrees (177 degrees C). In this measurement example, the target temperature changes depending on how close the Remote Sensing Thermometer is to the fire. This is because the fire's warmth is spread across a larger area at farther distances. You can feel this same effect on your hand if you hold your hand closer or farther from the fire—your hands feel warmer the closer you get—though don't get your hand too close to the fire!

FIGURE 12-6: The Remote Sensing Thermometer can measure the temperature of a fire at a safe distance.

the heat-seeking cannon

The Heat-Seeking Cannon, shown in Figure 12-7, scans an area, looking for a hot target at which to shoot. When the cannon locates this target, it activates two projectile launchers by using a motor that shoots 12 studs at the target. You'll program how hot the target should be to avoid unintended misfires. Only a target that is warmer than room temperature will trigger the projectile launchers. The IR Temperature Sensor will measure the potential target's temperature by using the same mount built earlier for the Remote Sensing Thermometer.

building the heat-seeking cannon

Use the following instructions to build the Heat-Seeking Cannon.

FIGURE 12-7: The Heat-Seeking Cannon scans an area for a hot target to shoot at.

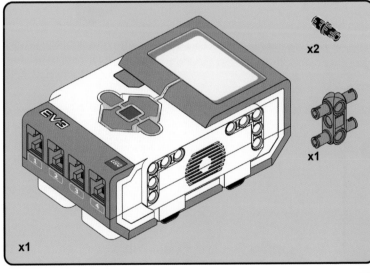

1

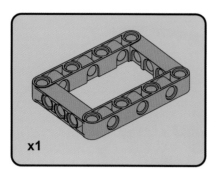

2

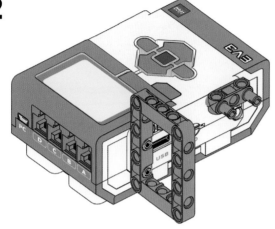

3

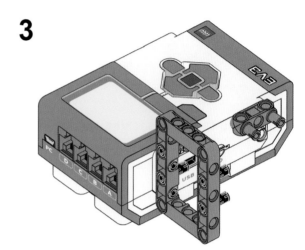

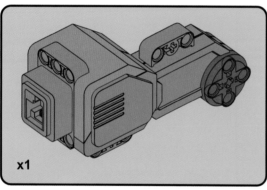

4

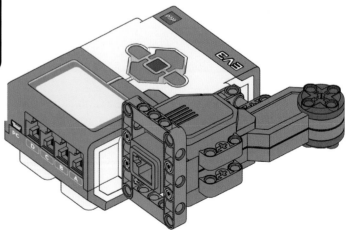

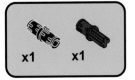

5

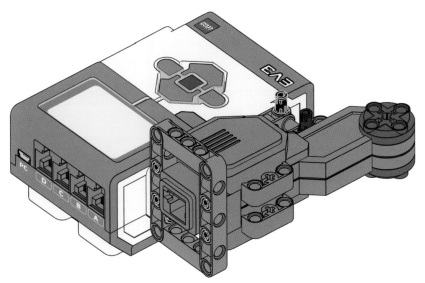

6

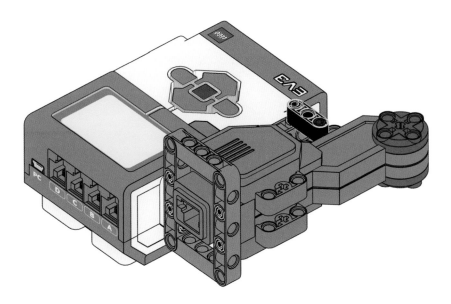

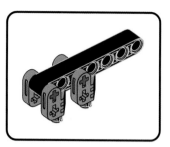

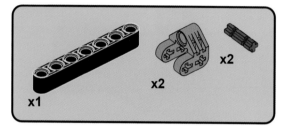

7

8

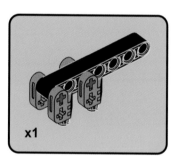

9

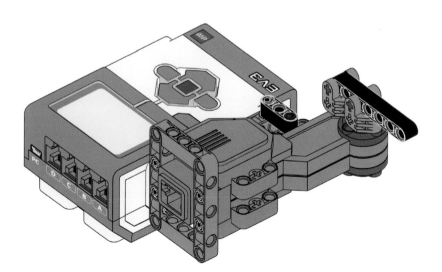

10

11

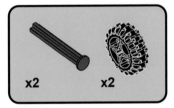

12

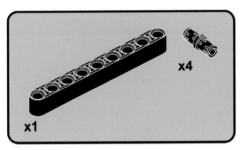

13

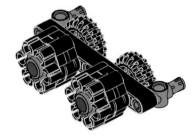

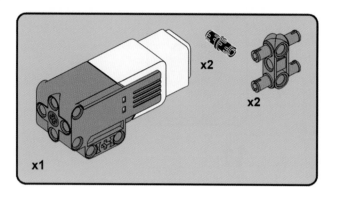

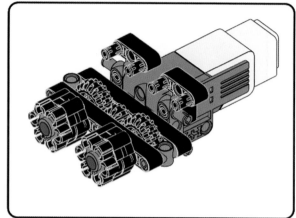

14

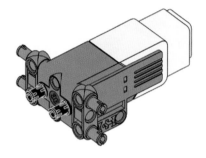

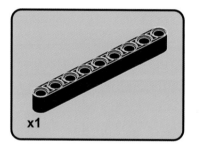

15

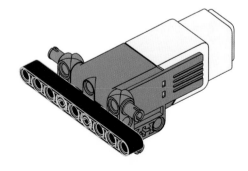

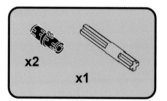

16

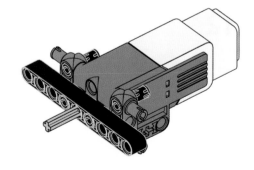

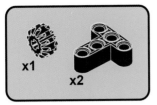

17

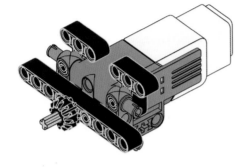

18

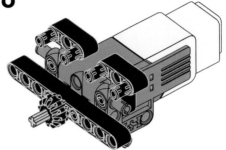

19

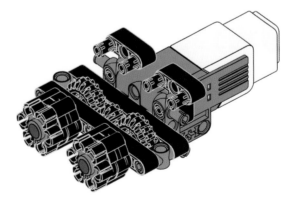

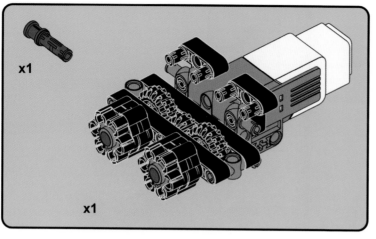

20

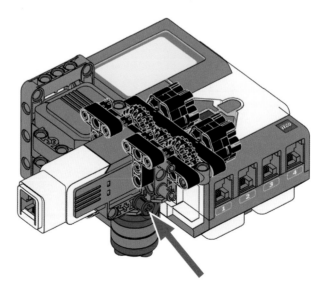

21

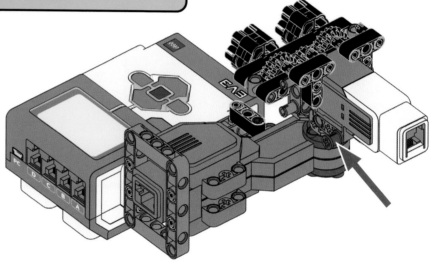

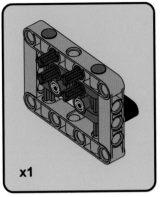

x1

22

Follow the instructions on
page 161 to build this mount for
the IR Temperature Sensor.

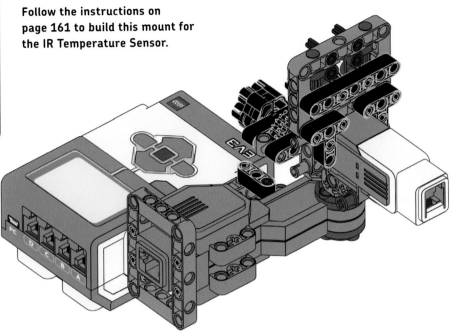

x12

23

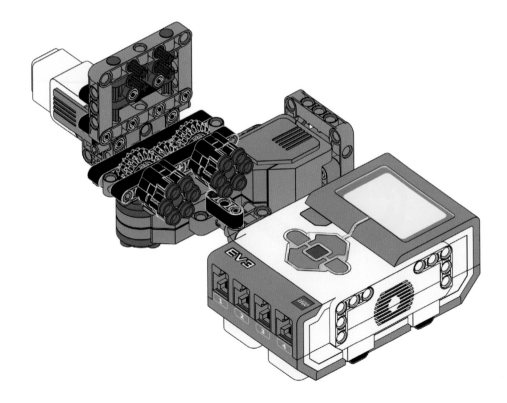

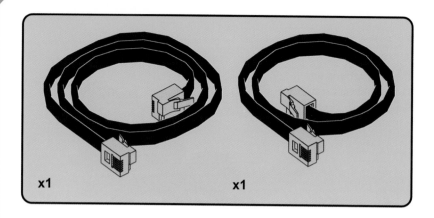

24

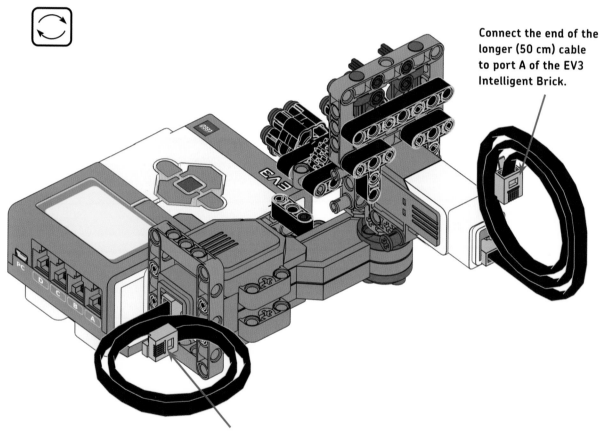

Connect the end of the longer (50 cm) cable to port A of the EV3 Intelligent Brick.

Connect the end of the shorter (35 cm) cable to port D of the EV3 Intelligent Brick.

Attach the Mindsensors IR Temperature Sensor upside-down onto the axles.

Plug the Mindsensors IR Temperature Sensor cable into port 1.

writing the software

The software for the Heat-Seeking Cannon, shown in Figure 12-8, runs two programs at the same time. One set of blocks reads measurements from the IR Temperature Sensor, displays the measurements on the front panel of the EV3 Intelligent Brick, and checks whether the target temperature is significantly higher than the ambient temperature.

If the target temperature is high enough, the software program activates the two projectile launchers by using an EV3 Medium Motor to shoot LEGO studs at the target.

The second program rotates the Heat-Seeking Cannon and scans the area to find a target. Once it finds and fires upon a target, the program stops, waiting for you to reload the projectile launchers.

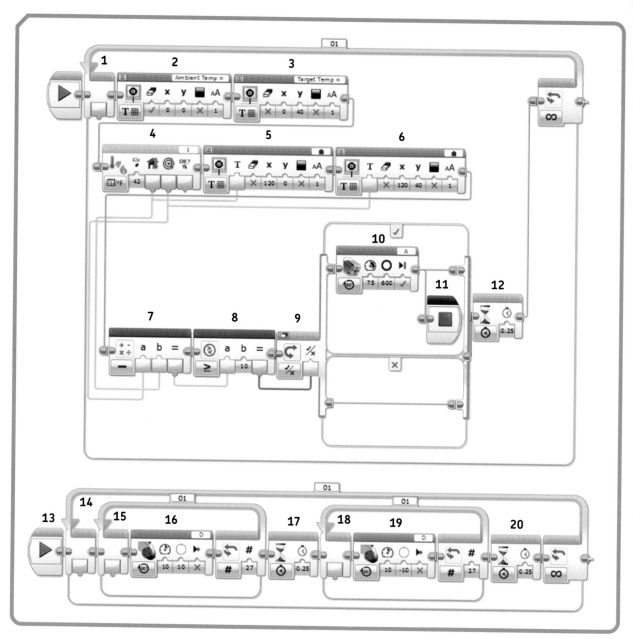

FIGURE 12-8: *The software for the Heat-Seeking Cannon uses two parallel programs.*

Use the following blocks to write the two programs, starting with the topmost program:

1. **Loop block**   This block infinitely repeats the temperature taking.

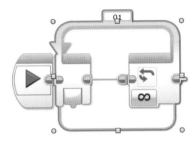

Place a Loop block after the program Start.

2. **Display block**   This prints a label on the front panel of the EV3 Intelligent Brick to label a measurement that is going to be made.

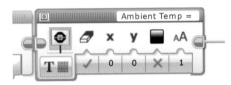

Insert a Display block inside the Loop. Set **Display** to **Text—Pixels**, **Text** to **Ambient Temp =**, **Clear Screen** to **True**, **x** to **0**, **y** to **0**, **Color** to **False**, and **Font** to **1**.

3. **Display block**   This prints another label on the front panel to label a second measurement.

Place a new Display block after the existing Display block. Set **Display** to **Text—Pixels**, **Text** to **Target Temp =**, **Clear Screen** to **False**, **x** to **0**, **y** to **40**, **Color** to **False**, and **Font** to **1**.

4. **Mindsensors IR Thermometer block**   This takes a measurement from the IR Temperature Sensor. You'll have to download this block from mindsensors.com and install it, after which point it should appear in the Sensor tab.

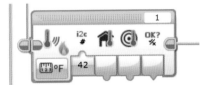

Place a Mindsensors IR Thermometer block after the Display block. Set **Mindsensors IR Thermometer** to **Measure-Fahrenheit** and **Port** to **1**.

5. **Display block**   This prints the ambient temperature measured by the IR Temperature Sensor on the front panel of the EV3 Intelligent Brick. The measurement will appear after the label programmed in block 2.

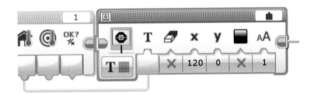

Place a Display block after the Mindsensors IR Thermometer block. Set **Display** to **Text—Pixels**, **Text** to **Wired**, **Clear Screen** to **False**, **x** to **120**, **y** to **0**, **Color** to **False**, and **Font** to **1**. Run a wire from the **Text** input to the **Ambient Fahrenheit Temperature** output on the Mindsensors IR Thermometer Block.

6. **Display block**   This prints the target temperature measured by the IR Temperature Sensor on the front panel of the EV3 Intelligent Brick. The measurement will appear after the label programmed in block 3.

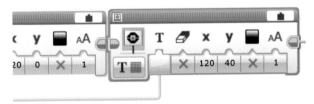

Place a new Display block after the previous Display block. Set **Display** to **Text—Pixels**, **Text** to **Wired**, **Clear Screen** to **False**, **x** to **120**, **y** to **40**, **Color** to **False**, and **Font** to **1**. Run a wire from **Text** input to the **Target Fahrenheit Temperature** output on the Mindsensors IR Thermometer Block.

7. **Math block**   This determines the difference between the target temperature and the ambient temperature.

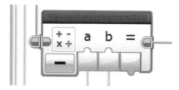

Place a Math block after the Display block. Set **Math** to **Subtract** and run a wire from the **a** input to the **Target Fahrenheit Temperature** output on the Mindsensors IR Thermometer block. Run another wire from the **b** input to the **Ambient Fahrenheit Temperature** output on the Mindsensors IR Thermometer block.

8. **Compare block**   This determines whether the target temperature is 10 degrees or more above the ambient temperature.

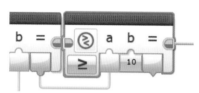

Place a Compare block after the Math block. Set **Compare** to **Greater Than or Equal To** and set **b** to **10**. Run a wire from the **a** input to the **Result** output on the Math block.

9. **Switch block**   This presents two options, based on the results of the Compare block, corresponding to whether or not the target temperature is significantly warmer than the ambient temperature. A true path is taken if the Compare block detects a hot target. Otherwise, the false path, in which no action takes place, executes.

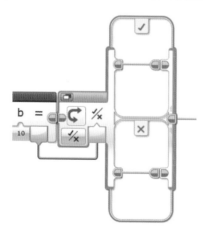

Place a Switch block after the Compare block. Set **Switch** to **Logic** and run a wire from the **Logic** input to the **Result** output on the Compare block.

10. **Medium Motor block**   This spins an EV3 Medium Motor that fires a pair of projectile launchers. This action takes place only when a hot target has been detected.

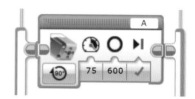

Place a Medium Motor block inside the true path of the Switch block. Set **Medium Motor** to **On for Degrees, Port** to **A**, **Power** to **75**, **Degrees** to **600**, and **Brake at End** to **True**.

11. **Stop Program block**   This halts execution of the program, stopping both temperature measurement and the parallel motor control program that enables the Heat-Seeking Cannon to scan in an angle. Because the Heat-Seeking Cannon has found a target and expended its ammunition, there is no need to continue the program.

Place a Stop Program block after the Medium Motor block. The EV3 programming environment will automatically create two connections after the Medium Motor block: one to the Stop Program block and a second to connect to the end of the Switch block's true path.

12. **Wait block**   This pauses action for a short period to prevent data from becoming confused as information is passed between the IR Temperature Sensor and the EV3 Intelligent Brick.

Place a Wait block after the Switch block. Set **Wait** to **Time** and **Seconds** to **0.25**.

13. **Start block**   This begins the second program, which will operate a motor that scans the target area.

Place this Start block in an open area of the work-space, underneath the set of commands for reading the IR Temperature Sensor. You can find the Start block under the Flow Control tab.

14. **Loop block**   This infinitely scans the angular view of the Heat-Seeking Cannon.

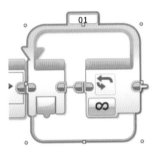

Place a Loop block after the Start block.

15. **Loop block**   This sets up a loop that will repeat 27 times. After this loop runs 27 times, it will stop, and the program will proceed to the next command.

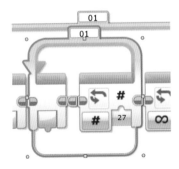

Place a Loop block inside the Loop block of step 14. Set **Loop** to **Count** and **Count** to **27**.

16. **Large Motor block**   This rotates the EV3 Large Motor by 10 degrees to increment the Heat–Seeking Cannon's angle. The program repeats this action every time the Loop block of step 15 runs. After the loop completes, the cannon will have scanned an area of 270 degrees. The EV3 Large Motor moves at a gentle power setting of 10 to avoid straining the cables that connect it to the EV3 Intelligent Brick.

Place a Large Motor block inside the count loop. Set **Large Motor** to **On for Degrees**, **Port** to **D**, **Power** to **10**, **Degrees** to **10**, and **Brake at End** to **False**.

17. **Wait block**   This pauses the action after the EV3 Large Motor has finished a scan. This delay keeps the motor from jerking before moving on to the next step in the program.

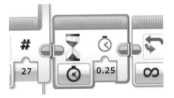

Place a Wait block after the Loop block. Set **Wait** to **Time** and **Seconds** to **0.25**.

18. **Loop block**   This loops 27 times, duplicating block 15. This new loop will return the motor scan angle to its starting point.

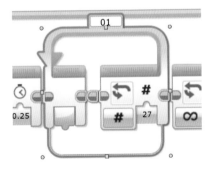

Place a Loop block after the Wait block. Set **Loop** to **Count** and **Count** to **27**.

19. **Large Motor block**   This rotates the EV3 Large Motor by –10 degrees. When the loop containing this motor action completes after 27 iterations, the EV3 Large Motor will be back to its starting position.

Place a Large Motor block inside the count loop. Set **Large Motor** to **On for Degrees**, **Port** to **D**, **Power** to **10**, **Degrees** to **–10**, and **Brake at End** to **False**.

20. **Wait block**   This pauses action after the EV3 Large Motor has finished a scan, as in step 17.

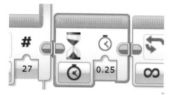

Place a Wait block after the Loop block. Set **Wait** to **Time** and **Seconds** to **0.25**.

using the heat-seeking cannon

Before activating the program, rotate the EV3 Medium Motor on its connection to the EV3 Large Motor so that the IR Temperature Sensor faces away from the EV3 Intelligent Brick, as pictured in Figure 12-9.

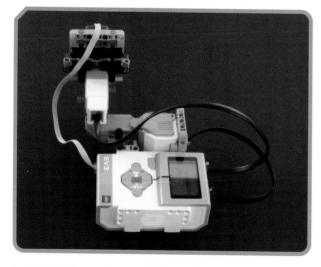

FIGURE 12-9: Manually turn the EV3 Medium Motor so that the IR Temperature Sensor faces away from the EV3 Intelligent Brick.

This sets the starting point for the cannon's scan of the environment. When you activate the program, the Heat-Seeking Cannon will rotate clockwise 270 degrees and then return to its starting position. Make sure that the connector cable plugged into the EV3 Medium Motor is long enough to accommodate the rotation. If the cable is too short, it will grow taut and rotation will stop. Arrange the cable so that it either lies flat, near the base of the EV3 Intelligent Brick, or forms a loop sticking upward that allows the cannon to pass underneath. Figure 12-9 shows the loop approach.

Now that you've adjusted the angles and positioned the connector cable, activate the software. The Heat-Seeking Cannon should begin scanning, looking for a hot target to shoot at. You can see the range of temperatures detected on the front panel of the EV3 Intelligent Brick. Now place a hot target in range of the Heat-Seeking Cannon—a cup of hot tea is a good target, as shown in Figure 12-10 and in a video at *http://hightechlego.com/*. If you would like to shoot at something less warm—say, your hand—lower the **b** setting of the Compare block in step 8 of the program. Just don't put your face in view of the Heat-Seeking Cannon!

FIGURE 12-10: A hot cup of tea serves as a good target for the Heat-Seeking Cannon.

what you learned

In this chapter, you learned to add a far-infrared radiometer to LEGO inventions to measure the surface temperature of an object without touching it. You used this technology to build a thermometer that points at objects and displays temperature measurements on the EV3 Intelligent Brick. In a second project, you built a heat-seeking weapon that shoots LEGO studs at hot targets. Finally, you learned an advanced programming technique to run two programs at the same time on the EV3 Intelligent Brick.

index

High-Tech LEGO Projects is set in Chevin. The book was printed and bound by Versa Printing, Inc. in East Peoria, Illinois. The paper is 70# White Coated (Matte).